Semiconductor Physics Modeling Made Easy

Jamie Flux

https://www.linkedin.com/company/golden-dawn-engineering/

Contents

1 Drude Model — 10
- Classical Electron Motion — 10
  - 1 Newton's Equations of Motion — 10
  - 2 Drude Equation — 11
- Electron Collisions — 11
  - 1 Scattering Mechanisms — 11
  - 2 Random Thermal Motion — 11
- Drude Conductivity — 11
- Limitations of the Drude Model — 12
  - 1 Failure to Explain Temperature Dependence — 12
  - 2 Neglect of Band Structure — 12
- Python Code Snippet — 12

2 Ohm's Law — 15
- Ohm's Law Formulation — 15
  - 1 Resistance and Resistivity — 15
  - 2 Temperature Dependence — 16
- Applicability to Semiconductor Physics — 16
  - 1 Non-Ohmic Behavior — 16
  - 2 Nonlinear Elements — 16
  - 3 Ohmic Contacts — 17
  - 4 Conditions for Ohmic Behavior — 17
- Conclusion — 17
- Python Code Snippet — 17

3 Quantum Mechanics Basics — 20
- Introduction to Quantum Mechanics — 20
  - 1 Wave-Particle Duality — 20
  - 2 Quantum Superposition — 21
  - 3 Schrödinger's Equation — 21

		The Time-Independent Schrödinger Equation	21
	1	Hamiltonian Operator	21
	2	Time-Independent Schrödinger Equation	22
	3	Boundary Conditions	22
		Application to Semiconductor Theory	22
	1	Effective Mass Approximation	22
	2	Energy Bands and Band Gaps	23
	3	Carrier Statistics	23
		Conclusion	23
		Python Code Snippet	23

4 Free Electron Model — 26

	Introduction	26
1	Assumptions	26
	Mathematical Formulation	27
1	Wave Function	27
2	Bloch's Theorem	27
3	Energy Bands	27
4	Fermi-Dirac Statistics	28
	Electronic Properties	28
1	Density of States	28
2	Effective Mass	28
	Conclusion	29
	Python Code Snippet	29

5 Energy Bands — 31

	Band Theory of Solids	31
1	Tight-Binding Approximation	31
2	Classification of Energy Bands	32
	Significance of Energy Bands	33
1	Electron Mobility	33
2	Electronic and Optical Properties	33
3	Thermal Conductivity	33
	Conclusion	34
	Python Code Snippet	34

6 Bloch's Theorem — 37

	Derivation of Bloch's Theorem	37
	Implications for Electrons in Periodic Potentials	38
1	Band Structure	38
2	Energy-Momentum Relationship	39
3	Density of States	39

    4   Band Gaps and Electronic Conductivity . . . 39
    Conclusion 39
    Python Code Snippet 40

7 Brillouin Zones   42
    Periodic Potentials and Reciprocal Space 42
    Brillouin Zones and the First Brillouin Zone 43
    Higher Brillouin Zones 43
    Importance in Band Theory 44
    Conclusion 44
    Python Code Snippet 45

8 Effective Mass Theorem   48
    Understanding Effective Mass 48
    Effective Mass Approximation 49
    Anisotropic Effective Mass 49
    Characteristics of Effective Mass 49
    Applications of Effective Mass 50
    Conclusion 51
    Python Code Snippet 51

9 Density of States   54
    Energy Bands and Energy States 54
    1   Energy Dispersion Relation 54
    2   Distribution of Energy States 55
    Derivation of Density of States 56
    Properties of the Density of States 56
    Conclusion 57
    Python Code Snippet 57

10 Fermi-Dirac Statistics   60
    Fermi-Dirac Distribution Function 60
    Occupation Probabilities in Semiconductors 61
    Importance in Semiconductor Physics 61
    Python Code Snippet 62

11 Carrier Concentration   65
    Intrinsic Carrier Concentration 65
    Extrinsic Carrier Concentration 66
    1   Law of Mass Action 66
    Impurity Ionization and Compensation 67
    1   Compensation 67
    Conclusion 67

    Python Code Snippet 68

12 Poisson's Equation   71
    Introduction to Poisson's Equation 71
    Role in Semiconductor Device Physics 71
    1   Poisson's Equation in pn Junctions 72
    2   Boundary Conditions 72
    Conclusion 73
    Python Code Snippet 73

13 Continuity Equation   77
    Derivation of the Continuity Equation 77
    1   Conservation of Charge 77
    2   Current Density and Carrier Motion 78
    3   Simplification and Final Form 79
    Implications and Applications 79
    Conclusion 79
    Python Code Snippet 80

14 Drift-Diffusion Model   82
    Drift and Diffusion 82
    1   Drift 82
    2   Diffusion 83
    Drift-Diffusion Equations 83
    Implications and Applications 84
    Python Code Snippet 85

15 Einstein Relation   88
    Derivation of the Einstein Relation 88
    Implications of the Einstein Relation 89
    Python Code Snippet 90

16 PN Junctions   92
    Depletion Approximation 92
    Built-in Potential 93
    Python Code Snippet 94

17 Shockley Equation   96
    Derivation of the Shockley Equation 96
    Implications and Applications 98
    Python Code Snippet 99

18 MOS Capacitor — 101

- Capacitance-Voltage Characteristics 101
  - 1 Basic Theory 101
  - 2 Depletion Region Capacitance 102
  - 3 Inversion Region Capacitance 102
  - 4 Oxide Capacitance 103
  - 5 Total Capacitance 103
- Surface Potential 103
  - 1 Basic Theory 103
  - 2 Flatband Voltage 104
  - 3 Threshold Voltage 104
  - 4 Subthreshold Swing 104
- Conclusion 105
- Python Code Snippet 105

19 Band Bending — 108

- Basic Theory and Governing Equations 108
  - 1 Poisson's Equation 108
  - 2 Space Charge Region 109
  - 3 Governing Equations 109
  - 4 Surface Potential 109
  - 5 Intrinsic Semiconductor 110
- Metal-Semiconductor Junctions 110
  - 1 Schottky Barrier 110
  - 2 PN Junctions 111
- Conclusion 111
- Python Code Snippet 111

20 Quantum Tunneling — 114

- Introduction to Quantum Tunneling 114
  - 1 Barrier Penetration Probability 114
- Impact on Semiconductor Devices 115
  - 1 Tunneling in Transistors 115
  - 2 Tunneling in Diodes 115
  - 3 Quantum Well Structures 115
  - 4 Tunneling Transistors 115
- Conclusion 116
- Python Code Snippet 116

21 Charge Control Model — 118
- Introduction to Charge Control Model 118
  - 1 Metal-Oxide-Semiconductor Structure 118
  - 2 Threshold Voltage 119
  - 3 Transconductance 119
- Conclusion 119
- Python Code Snippet 120

22 Small-Signal Model — 122
- Introduction to Small-Signal Model 122
  - 1 Linearization 122
  - 2 Small-Signal Equivalent Circuit 123
- Small-Signal Analysis of Semiconductors 123
  - 1 Small-Signal Gain 123
  - 2 Input and Output Impedances 123
  - 3 Frequency Response 123
- Conclusion 124
- Python Code Snippet 124

23 Boltzmann Transport Equation — 127
- Introduction to the Boltzmann Transport Equation . 127
  - 1 Distribution Function 128
  - 2 Carrier Velocity 128
  - 3 Collision Term 128
- Solution Techniques for the Boltzmann Transport Equation 128
  - 1 Relaxation Time Approximation 129
  - 2 Monte Carlo Simulation 129
  - 3 Numerical Methods 129
- Applications of the Boltzmann Transport Equation . 130
- Conclusion 130
- Python Code Snippet 130

24 Heterostructures — 133
- Energy Band Alignment in Heterostructures 133
  - 1 Type I Band Alignment 134
  - 2 Type II Band Alignment 134
- Transport Properties in Heterostructures 135
  - 1 Tunneling 135
  - 2 Carrier Confinement 135
- Conclusion 136
- Python Code Snippet 136

25 Quantum Confinement — 139
- Energy Quantization in Quantum Wells 139
- Density of States in Quantum Wells 140
- Carrier Confinement and Wavefunctions 141
- Conclusion 141
- Python Code Snippet 142

26 Thermal Effects — 144
- Thermal Conductivity and Heat Transfer 144
- Temperature Dependence of Thermal Conductivity . 145
- Impact on Semiconductor Behavior 145
- Thermal Conductivity Measurements 146
- Conclusion 147
- Python Code Snippet 147

27 Noise Models — 149
- Sources of Noise in Semiconductor Devices 149
  - 1 Thermal Noise 149
  - 2 Shot Noise 150
  - 3 Flicker or 1/f Noise 150
  - 4 Quantum Noise 150
- Noise Models 150
  - 1 Power Spectral Density (PSD) 151
  - 2 Nyquist Formula 151
  - 3 White Noise Model 151
  - 4 Correlation Function 151
  - 5 Noise Figure 151
- Conclusion 152
- Python Code Snippet 152

28 Carrier Recombination-Generation — 155
- 1 Introduction 155
- 2 Rate Equations 155
- 3 Detailed Balance Equation 156
- 4 Recombination Mechanisms 156
- 5 Generation Mechanisms 157
- 6 Impacts on Semiconductor Performance . . . 158
- Python Code Snippet 158

29 Non-Equilibrium Green's Function (NEGF)   161
    1   Introduction to NEGF 161
    2   Fundamentals of NEGF 162
    3   Dyson's Equation 162
    4   Electron and Hole Propagation 162
    5   Self-Energy 162
    6   Transport Equations 162
    7   Landauer-Büttiker Formalism 163
    8   Applications of NEGF 163
    9   Challenges and Future Directions 163
    Python Code Snippet 164

30 Phonon Interaction   166
    1   Introduction to Phonons 166
    2   Lattice Dynamics 166
    3   Normal Modes 167
    4   Phonon Dispersion 167
    5   Phonon-Phonon Interactions 167
    6   Thermal Conductivity 167
    7   Phonons and Carrier Mobility 167
    8   Phonon Drag 168
    9   Thermoelectric Effects 168
    10   Phonon Engineering 168
    Python Code Snippet 169

31 Spintronics   171
    Electron Spin 171
    Spin-Orbit Interaction 171
    Spin-Dependent Transport 172
    Spin Relaxation 172
    Spin Hall Effect 172
    Spin-Based Devices 172
    Spin-Orbit Torque 173
    Magnetization Dynamics 173
    Spin-Photon Interaction 173
    Python Code Snippet 174

32 Optoelectronics   177
    Optical Absorption 177
    Direct and Indirect Band Gap Semiconductors . . . 178
    Recombination and Generation Rates 178
    Radiative and Non-Radiative Recombination 178

    Luminescence and Electroluminescence 179
    Theoretical Models for Light-Matter Interaction . . . 179
    Photonic Structures 179
    Optical Waveguides 180
    Photodetectors 180
    Solar Cells 180
    Python Code Snippet 181

33 Advanced Computational Methods   184
    Finite Element Method (FEM) 184
    Mesh Generation 185
    Time-Dependent Simulations 185
    Carrier Transport Models 186
    Schrodinger-Poisson Solver 186
    Other Numerical Approaches 186
    Software and Simulation Tools 187
    Parallel Computing 187
    Model Validation and Calibration 187
    Python Code Snippet 188

Chapter 1

Drude Model

In this chapter, we explore the classical Drude model of electrical conduction in metals. The Drude model is a simplified yet valuable representation of the behavior of electrons in a metal under the influence of an electric field.

Classical Electron Motion

In the Drude model, we consider the motion of free electrons in a metal as classical particles. Although electrons are fundamentally quantum mechanical particles, the Drude model neglects their wave-particle duality and treats them purely as classical point particles.

1 Newton's Equations of Motion

The motion of an individual electron in an electric field is described by Newton's second law of motion:

$$m\frac{d^2\mathbf{r}}{dt^2} = -e\mathbf{E}$$

where m is the mass of the electron, $\mathbf{r}$ is the position vector of the electron, e is the charge of the electron, and $\mathbf{E}$ is the applied electric field.

2 Drude Equation

To describe the collective behavior of a large number of electrons, we introduce the concept of electron density, n, which represents the number of free electrons per unit volume. Assuming that the thermal motion of electrons can be neglected, we can write the equation of motion for the average velocity of the electrons, **v**, as follows:

$$\frac{d\mathbf{v}}{dt} = -\frac{e\mathbf{E}}{m} - \frac{\mathbf{v}}{\tau}$$

where τ is the relaxation time, which characterizes the average time between electron collisions.

Electron Collisions

In the Drude model, electron collisions are introduced as a simplified mechanism to account for the scattering and randomization of electron motion within the metal. These collisions are modeled using a phenomenological relaxation time, τ, which is characteristic of the material and the temperature at which it is operating.

1 Scattering Mechanisms

Electrons can experience various scattering mechanisms in metals, such as impurity scattering, lattice scattering, and electron-electron scattering. These mechanisms can be characterized by an average scattering time, τ_s, which represents the average time between scattering events.

2 Random Thermal Motion

At finite temperatures, electrons exhibit random thermal motion, which introduces additional scattering and chaotic behavior. This motion is neglected in the Drude model, assuming the electrons are at absolute zero temperature.

Drude Conductivity

The Drude model provides a simple relationship between the electric field and the resulting current in a metal, represented by the

electrical conductivity, σ. The equation for Drude conductivity is given by:
$$\mathbf{J} = \sigma \mathbf{E}$$
where **J** is the current density in the metal.

Limitations of the Drude Model

While the Drude model is a useful starting point for understanding electrical conduction in metals, it has several limitations. It neglects quantum effects, such as wave-particle duality and quantum tunneling, which become significant at small scales. Additionally, the Drude model assumes a time-independent electric field and a perfectly regular lattice structure, which may not accurately represent real-world materials.

1 Failure to Explain Temperature Dependence

One major limitation of the Drude model is its inability to explain the temperature dependence of electrical conductivity in metals, particularly at low temperatures. Experimental observations clearly demonstrate that conductivity decreases as the temperature approaches absolute zero, which contradicts the assumption of a constant relaxation time in the Drude model.

2 Neglect of Band Structure

The Drude model oversimplifies the behavior of electrons by neglecting the band structure of the metal. In reality, the presence of energy bands and band gaps greatly influences the electronic properties of materials, especially in semiconductors and insulators. This omission limits the applicability of the Drude model to metals only.

In the next chapter, we will introduce Ohm's Law and its mathematical formulation, which builds upon the concepts established in the Drude model.

Python Code Snippet

Below is a Python code snippet that calculates the average velocity of electrons, the current density using the Drude model, and evaluates the electrical conductivity in metals based on the given parameters.

```python
def calculate_average_velocity(electric_field, relaxation_time):
    '''
    Calculate the average velocity of electrons in the presence of
      an electric field.
    :param electric_field: The applied electric field in volts per
      meter (V/m).
    :param relaxation_time: The relaxation time in seconds (s).
    :return: Average velocity of electrons in meters per second
      (m/s).
    '''
    charge = 1.602e-19   # Charge of an electron in coulombs
    mass = 9.109e-31     # Mass of an electron in kg
    average_velocity = (charge * electric_field * relaxation_time) /
      mass
    return average_velocity


def calculate_current_density(electric_field, conductivity):
    '''
    Calculate the current density based on the electric field and
      electrical conductivity.
    :param electric_field: The applied electric field in volts per
      meter (V/m).
    :param conductivity: The electrical conductivity in siemens per
      meter (S/m).
    :return: Current density in amperes per square meter (A/m^2).
    '''
    current_density = conductivity * electric_field
    return current_density


def calculate_conductivity(carrier_density, charge, mobility):
    '''
    Calculate the electrical conductivity based on carrier density,
      charge, and mobility.
    :param carrier_density: The number of carriers per unit volume
      in m^-3.
    :param charge: Charge of the carriers in coulombs.
    :param mobility: Mobility of the carriers in m^2/(V*s).
    :return: Electrical conductivity in siemens per meter (S/m).
    '''
    conductivity = carrier_density * charge * mobility
    return conductivity


# Constants and inputs for calculations
electric_field = 1000    # Electric field in volts per meter (V/m)
relaxation_time = 1e-14  # Relaxation time in seconds (s)
carrier_density = 1e28   # Carrier density in m^-3
```

```
charge = 1.602e-19   # Charge of an electron in coulombs
mobility = 0.14   # Mobility of carriers in m^2/(V*s)

# Calculations
average_velocity = calculate_average_velocity(electric_field,
↪   relaxation_time)
current_density = calculate_current_density(electric_field,
↪   calculate_conductivity(carrier_density, charge, mobility))
conductivity = calculate_conductivity(carrier_density, charge,
↪   mobility)

# Output results
print("Average Velocity of Electrons:", average_velocity, "m/s")
print("Current Density:", current_density, "A/m^2")
print("Electrical Conductivity:", conductivity, "S/m")
```

This code defines three functions:

- calculate_average_velocity computes the average velocity of electrons under an electric field using relaxation time.
- calculate_current_density calculates the current density based on the applied electric field and electrical conductivity.
- calculate_conductivity determining the electrical conductivity using carrier density, charge, and mobility.

The provided example calculates the average velocity of electrons, the resulting current density, and the electrical conductivity in a metal, then prints the results.

Chapter 2

Ohm's Law

In this chapter, we delve into the mathematical formulation of Ohm's Law and explore its applicability to semiconductor physics. Ohm's Law is a fundamental principle in electrical circuit analysis that relates the current flowing through a conductor to the voltage across it. It provides valuable insights into the behavior of conductive materials and can be applied to understand the electrical properties of semiconductors.

Ohm's Law Formulation

Ohm's Law is typically stated as follows: the current flowing through a conductor is directly proportional to the voltage across it, and the constant of proportionality is the conductance. Mathematically, Ohm's Law can be expressed as:

$$I = \frac{V}{R}$$

where I represents the electric current in amperes (A), V denotes the voltage in volts (V), and R represents the resistance in ohms (Ω).

1 Resistance and Resistivity

Resistance, denoted by R, quantifies the opposition that a material offers to the flow of electric current. It is related to the resistivity of the material, ρ, by the formula:

$$R = \rho \cdot \frac{L}{A}$$

where L is the length of the conductor and A is its cross-sectional area.

2 Temperature Dependence

It is important to note that the resistance of a material can be temperature-dependent. For many metallic conductors, resistance increases with temperature, following a linear relationship. This temperature dependence can be described by the temperature coefficient of resistance, α, such that:

$$\Delta R = R_0 \cdot \alpha \cdot \Delta T$$

where ΔR is the change in resistance, R_0 is the resistance at a reference temperature, α is the temperature coefficient of resistance, and ΔT is the change in temperature.

Applicability to Semiconductor Physics

While Ohm's Law is widely utilized in traditional electrical circuit analysis, its application to semiconductor physics requires consideration of additional factors.

1 Non-Ohmic Behavior

In semiconductors, the relationship between current and voltage is not always linear or governed solely by resistance. This departure from Ohm's Law is known as non-Ohmic behavior. Non-Ohmic behavior often arises due to the influence of carrier lifetime, doping concentration, and temperature on carrier mobility.

2 Nonlinear Elements

The behavior of semiconductor devices, such as diodes and transistors, cannot be accurately described by Ohm's Law alone. These devices exhibit highly nonlinear characteristics and require more complex mathematical models to analyze their performance.

3 Ohmic Contacts

Ohmic contacts, also referred to as Ohmic interfaces, are essential for establishing good electrical connections to semiconductors. In these contacts, the current-voltage relationship follows Ohm's Law since the voltage drop primarily occurs across the contact itself rather than the semiconductor material.

4 Conditions for Ohmic Behavior

In some cases, semiconductors can exhibit Ohmic behavior under certain conditions. This occurs when the concentration and mobility of majority carriers dominate the transport properties, and the injection and recombination of minority carriers are negligible.

Conclusion

In this chapter, we explored the mathematical formulation of Ohm's Law and its applicability to semiconductor physics. Although Ohm's Law provides a fundamental understanding of the relationship between current, voltage, and resistance, its straightforward application is limited in the complex realm of semiconductor devices and materials. Development of more advanced models and analysis techniques is necessary to fully comprehend and predict the behavior of semiconductors.

Python Code Snippet

Below is a Python code snippet that implements key equations and algorithms discussed in this chapter regarding Ohm's Law and resistance.

```python
def calculate_current(voltage, resistance):
    '''
    Calculate the electric current using Ohm's Law.
    :param voltage: Voltage in volts (V).
    :param resistance: Resistance in ohms ().
    :return: Current in amperes (A).
    '''
    if resistance == 0:
        raise ValueError("Resistance cannot be zero.")
    return voltage / resistance
```

```python
def calculate_resistance(resistivity, length, area):
    '''
    Calculate resistance using the resistivity, length, and
    ↪ cross-sectional area of the conductor.
    :param resistivity: Resistivity of the material in ohm-meters
    ↪ (m).
    :param length: Length of the conductor in meters (m).
    :param area: Cross-sectional area of the conductor in square
    ↪ meters (m²).
    :return: Resistance in ohms ().
    '''
    return resistivity * (length / area)

def temperature_coefficient_of_resistance(R0, alpha, delta_T):
    '''
    Calculate the change in resistance due to temperature change.
    :param R0: Resistance at reference temperature in ohms ().
    :param alpha: Temperature coefficient of resistance.
    :param delta_T: Change in temperature in degrees Celsius (°C).
    :return: Change in resistance in ohms ().
    '''
    return R0 * alpha * delta_T

# Inputs for the calculations
voltage = 10.0   # Voltage in volts
resistance = 5.0   # Resistance in ohms
resistivity = 1.68e-8   # Resistivity of copper in ohm-meters (m)
length = 2.0   # Length of wire in meters
cross_section_area = 1.0e-6   # Cross-sectional area in square meters
↪ (m·)
R0 = 5.0   # Resistance at reference temperature
alpha = 0.0039   # Temperature coefficient of resistance for copper
delta_T = 25.0   # Change in temperature

# Calculations
current = calculate_current(voltage, resistance)
calculated_resistance = calculate_resistance(resistivity, length,
↪ cross_section_area)
resistance_change = temperature_coefficient_of_resistance(R0, alpha,
↪ delta_T)

# Output results
print("Calculated Current:", current, "A")
print("Calculated Resistance:", calculated_resistance, "")
print("Change in Resistance due to Temperature:", resistance_change,
↪ "")
```

This code defines three functions:

- `calculate_current` computes the electric current based on voltage and resistance using Ohm's Law.

- `calculate_resistance` calculates the resistance of a conductor based on its resistivity, length, and area.
- `temperature_coefficient_of_resistance` determines the change in resistance due to variations in temperature.

The provided example performs calculations for electric current, resistance, and temperature effects on resistance, then prints the results.

Chapter 3

Quantum Mechanics Basics

In this chapter, we will provide a comprehensive review of Schrödinger's equation and its application to semiconductor theory. Quantum mechanics forms the foundation for understanding the behavior of electrons and other particles at the atomic and subatomic levels. By delving into the mathematical framework of Schrödinger's equation, we will gain insights into the wave-particle duality of matter and its significance in semiconductor physics.

Introduction to Quantum Mechanics

Quantum mechanics is a branch of physics that was developed in the early 20th century to explain the peculiar behavior of particles on the atomic scale. It superseded classical mechanics, which had successfully described the motion of macroscopic objects. The principles of quantum mechanics allow for the description and prediction of a particle's behavior both as a wave and as a particle.

1 Wave-Particle Duality

One of the fundamental concepts of quantum mechanics is the wave-particle duality. It suggests that particles, such as electrons, exhibit both wave-like and particle-like properties depending on the experimental context. This duality is mathematically represented by wave functions, which describe the probability distribution of finding a particle at a given position.

2 Quantum Superposition

Quantum mechanics permits the idea of superposition, where particles can exist in multiple states simultaneously. This concept defies classical intuition but has been experimentally confirmed. Superposition allows for the interference of wave functions, leading to phenomena such as diffraction and interference patterns.

3 Schrödinger's Equation

Schrödinger's equation is the mathematical framework that describes the behavior of quantum mechanical systems. In its time-independent form, the equation is written as follows:

$$\hat{H}\Psi = E\Psi$$

where $\hat{H}$ is the Hamiltonian operator, Ψ is the wave function, and E is the energy eigenvalue. Solving Schrödinger's equation provides information about the wave function and allows us to determine the energy spectrum of a quantum system.

The Time-Independent Schrödinger Equation

The time-independent Schrödinger equation plays a fundamental role in the study of semiconductor physics. By considering a particle in a potential energy field, we can derive the time-independent Schrödinger equation using the principles of classical mechanics and wave-particle duality.

1 Hamiltonian Operator

The Hamiltonian operator, denoted as $\hat{H}$, represents the total energy of a system. It is a sum of the kinetic energy operator, $\hat{T}$, and the potential energy operator, $\hat{V}$:

$$\hat{H} = \hat{T} + \hat{V}$$

The kinetic energy operator is given by:

$$\hat{T} = -\frac{\hbar^2}{2m}\nabla^2$$

where $\hbar$ is the reduced Planck's constant, m is the mass of the particle, and ∇^2 is the Laplacian operator. The potential energy operator, $\hat{V}$, depends on the specific system under consideration.

2 Time-Independent Schrödinger Equation

For a particle in a potential well, the time-independent Schrödinger equation is given by:

$$\hat{H}\Psi = E\Psi$$

where Ψ represents the wave function, and E is the corresponding energy eigenvalue. This equation can be solved to obtain the wave function and energy levels of the system.

3 Boundary Conditions

Solving the time-independent Schrödinger equation requires the application of appropriate boundary conditions. These conditions depend on the specific physical system and are necessary for obtaining physically meaningful solutions. Common boundary conditions include continuity of the wave function and the requirement for the wave function to approach zero at infinity.

Application to Semiconductor Theory

The time-independent Schrödinger equation finds wide application in semiconductor theory. By solving this equation for the electrons in a crystalline lattice, we can understand the electronic structure and energy band formation in semiconductors. The solutions to the Schrödinger equation provide insights into fundamental properties such as electron wave functions, energy levels, and band structures.

1 Effective Mass Approximation

In the context of semiconductors, the effective mass approximation is often employed to simplify the time-independent Schrödinger equation. This approximation treats the behavior of electrons as if they were free particles with an effective mass. By assuming a parabolic band structure near the band edges, the equation can be simplified to a form that is easier to solve.

2 Energy Bands and Band Gaps

Solving the time-independent Schrödinger equation for electrons in a crystalline lattice reveals the formation of energy bands. Energy bands are ranges of allowed energy levels for electrons within the crystal. These bands are separated by band gaps, where no electronic states exist. In semiconductor materials, the size of the band gap determines their classification as either intrinsic (with a sizable band gap) or extrinsic (with a smaller or no band gap).

3 Carrier Statistics

The solutions to the time-independent Schrödinger equation also provide a basis for understanding carrier statistics in semiconductors. The wave functions of the electrons yield the probability densities and distribution of electrons within the energy bands. This information is crucial for determining carrier concentration and intrinsic properties such as Fermi-Dirac statistics.

Conclusion

In this chapter, we explored the foundations of quantum mechanics and its application to semiconductor theory. Schrödinger's equation serves as a powerful mathematical tool for understanding the behavior of subatomic particles and provides valuable insights into the electronic properties of semiconductor materials. By solving the time-independent Schrödinger equation, we can determine wave functions, energy levels, and band structures, thus paving the way for further analysis and investigation in the field of semiconductor physics.

Python Code Snippet

Below is a Python code snippet that demonstrates how to solve the time-independent Schrödinger equation for a particle in a one-dimensional potential well, calculates energy levels, and determines the wave function for a specified quantum state.

```python
import numpy as np
import matplotlib.pyplot as plt

def solve_schrodinger_1d(potential, x_range, n):
    '''
    Solve the time-independent Schrödinger equation for a particle
        in a 1D potential.
    :param potential: Function representing the potential energy as
        a function of position x.
    :param x_range: Tuple representing the range of x values (x_min,
        x_max).
    :param n: Quantum number representing the state to solve for.
    :return: Tuple of (x, wave_function, energy).
    '''
    x_min, x_max = x_range
    x = np.linspace(x_min, x_max, 1000)

    # Set up the Hamiltonian operator as a second derivative matrix
    h = (x_max - x_min) / 1000  # Step size
    H = -0.5 * np.diag(np.ones(len(x)-1), -1) + \
        np.diag(np.ones(len(x)-1), 1) / (h**2) + \
        np.diag(potential(x) + np.zeros(len(x)))  # Include
        # potential in the diagonal
    )

    # Solve the eigenvalue problem
    energies, wave_functions = np.linalg.eigh(H)

    # Return the n-th energy and corresponding wave function
    return x, wave_functions[:, n], energies[n]

def well_potential(x):
    '''
    Potential energy function for an infinite square well.
    :param x: Position in units of length.
    :return: Potential energy in Joules.
    '''
    V0 = 1e-18  # Potential energy outside the well
    L = 1e-9   # Length of the well in meters
    return np.where((x < 0) | (x > L), V0, 0)  # 0 inside, V0
        # outside

# Define parameters for the simulation
x_range = (0, 2e-9)  # 2-nanometer wide well
n = 0  # Ground state

# Solve Schrödinger equation
x, wave_function, energy = solve_schrodinger_1d(well_potential,
    x_range, n)

# Normalize wave function
wave_function /= np.max(np.abs(wave_function))
```

```
# Output results
print(f"Energy of state {n}: {energy:.2e} J")
print(f"Wave Function (normalized) at x = {x}: {wave_function}")

# Plotting the results
plt.figure(figsize=(10, 5))
plt.plot(x, wave_function**2, label='Probability Density',
    color='blue')
plt.title('Probability Density for Ground State in Infinite Square
    Well')
plt.xlabel('Position (m)')
plt.ylabel('Probability Density')
plt.grid()
plt.legend()
plt.show()
```

This code defines two primary functions:

- `solve_schrodinger_1d` solves the time-independent Schrödinger equation for a specified potential and quantum state, returning the position values, normalized wave function, and energy for that state.
- `well_potential` returns the potential energy for a one-dimensional infinite square well.

The provided example computes the wave function and energy level for the ground state of a particle in an infinite square well and visualizes the probability density associated with the wave function.

Chapter 4

Free Electron Model

The free electron model is a simplified yet powerful approach to understanding the behavior of electrons in conducting materials. This model, also known as the free electron gas model, provides valuable insights into the electronic properties and transport phenomena in metals and other conductors. In this chapter, we will explore the fundamental principles behind the free electron model and its mathematical formulation.

Introduction

The free electron model considers an ensemble of electrons in a solid material, neglecting their interactions with other electrons and lattice vibrations. This approximation assumes that the interaction potential between electrons is sufficiently screened such that their individual motions can be treated independently.

1 Assumptions

The free electron model is based on several key assumptions:

1. The electrons are considered as non-interacting particles.

2. The electrons move in a periodic potential created by the lattice of the material.

3. The effects of external fields, such as electromagnetic fields, are included to describe the electron's behavior under various conditions.

With these assumptions, the free electron model allows for the analysis of the electronic structure and transport properties of conducting materials.

Mathematical Formulation

The free electron model employs quantum mechanics to describe the behavior of electrons in a crystalline lattice. By applying the principles of wave-particle duality and quantum statistics, we can derive certain fundamental concepts within this model.

1 Wave Function

The wave function, denoted as $\Psi(\mathbf{r}, t)$, describes the probability amplitude for finding an electron at position $\mathbf{r}$ at time t. In the free electron model, the wave function satisfies the time-dependent Schrödinger equation:

$$i\hbar \frac{\partial \Psi(\mathbf{r}, t)}{\partial t} = \left(-\frac{\hbar^2}{2m} \nabla^2 + V(\mathbf{r}, t) \right) \Psi(\mathbf{r}, t)$$

where $\hbar$ is the reduced Planck's constant, m is the mass of the electron, and $V(\mathbf{r}, t)$ is the potential energy due to the periodic lattice.

2 Bloch's Theorem

Bloch's theorem states that the solutions to the Schrödinger equation in a periodic potential can be expressed as the product of a plane wave and a periodic function. Mathematically, this can be written as:

$$\Psi(\mathbf{r}, t) = e^{i\mathbf{k}\cdot\mathbf{r}} u(\mathbf{r}, t)$$

where $e^{i\mathbf{k}\cdot\mathbf{r}}$ represents the plane wave part and $u(\mathbf{r}, t)$ is a periodic function with the same periodicity as the lattice potential.

3 Energy Bands

Solving the Schrödinger equation with the Bloch form of the wave function leads to the concept of energy bands. In the free electron model, these bands correspond to ranges of allowed energies for the electrons. The energy bands are separated by energy gaps where electrons cannot exist.

4 Fermi-Dirac Statistics

Within the free electron model, the electrons are treated as indistinguishable particles subject to Fermi-Dirac statistics. This statistical distribution determines the probability of finding an electron with a specific energy level in thermal equilibrium at a given temperature T. The Fermi-Dirac distribution function is defined as:

$$f(E) = \frac{1}{1 + e^{(E-\mu)/k_B T}}$$

where E is the energy level, μ is the chemical potential, k_B is the Boltzmann constant, and T is the temperature. The chemical potential represents the energy level at which the probability of occupation is 50

Electronic Properties

The free electron model provides insights into various electronic properties of conducting materials. These properties arise from the behavior of electrons within the crystalline lattice.

1 Density of States

The density of states (DOS) describes the distribution of allowed energy states for electrons in the material. In the free electron model, the DOS can be given by:

$$D(E) = \frac{V}{4\pi^2} \left(\frac{2m}{\hbar^2}\right)^{3/2} \sqrt{E}$$

where V is the volume of the material and E is the energy level. The DOS determines the number of available states per unit energy range.

2 Effective Mass

The effective mass, denoted as m^*, characterizes the behavior of electrons in the crystal lattice under the influence of the periodic potential. It represents an effective inertia that relates the acceleration of an electron to the applied force. Within the free electron model, the effective mass is constant and independent of energy.

Conclusion

The free electron model provides a valuable starting point for understanding the electronic properties and behaviors of conducting materials. By neglecting electron-electron interactions and accounting for the periodic potential, this model successfully captures key features such as energy bands, Fermi-Dirac statistics, and the density of states. The free electron model lays the foundation for further analysis of transport phenomena and the development of more sophisticated models in semiconductor physics.

Python Code Snippet

Below is a Python code snippet that implements key equations and algorithms related to the free electron model, including the density of states, effective mass, Fermi-Dirac statistics, and the wave function properly wrapped in LaTeX using the minted package.

```python
import numpy as np

def density_of_states(E, V, m):
    '''
    Calculate the density of states for free electrons.
    :param E: Energy level in joules.
    :param V: Volume of the material in cubic meters.
    :param m: Electron mass in kg.
    :return: Density of states in states per cubic meter per joule.
    '''
    return (V / (4 * np.pi**2)) * ((2 * m / (hbar**2))**(3/2) *
        np.sqrt(E)

def fermi_dirac_distribution(E, mu, T):
    '''
    Calculate the Fermi-Dirac distribution function.
    :param E: Energy level in joules.
    :param mu: Chemical potential in joules.
    :param T: Temperature in Kelvin.
    :return: Probability of occupation at energy E.
    '''
    k_B = 1.380649e-23  # Boltzmann constant in J/K
    return 1 / (1 + np.exp((E - mu) / (k_B * T)))

def wave_function(k, r):
    '''
    Calculate the wave function for free electrons.
    :param k: Wave vector in reciprocal meters.
    :param r: Position vector in meters.
```

```python
    :return: Value of the wave function.
    '''
    return np.exp(1j * np.dot(k, r))

# Constants
hbar = 1.0545718e-34  # Reduced Planck's constant in J.s
m = 9.10938356e-31  # Mass of the electron in kg
V = 1e-28  # Volume in cubic meters (example volume)
T = 300  # Temperature in Kelvin (room temperature)
mu = 5.1e-19  # Chemical potential in joules (example value)

# Energy range for density of states calculation
energy_levels = np.linspace(0, 1e-18, 100)  # Energy levels in
 ↪ joules
dos_values = [density_of_states(E, V, m) for E in energy_levels]

# Example input for Fermi-Dirac distribution
E_example = 1.6e-19  # Energy in joules (example energy)
fd_value = fermi_dirac_distribution(E_example, mu, T)

# Print results
print("Density of States Values:")
for E, dos in zip(energy_levels, dos_values):
    print(f"Energy: {E:.2e} J, DOS: {dos:.2e} states/m^3/J")

print(f"\nFermi-Dirac Distribution at E = {E_example:.2e} J:
 ↪ {fd_value:.4f}")

# Example position and wave vector for wave function calculation
k_example = np.array([1, 1, 1]) * (2 * np.pi / 1e-9)  # example wave
 ↪ vector
r_example = np.array([0, 0, 0])  # position vector
wf_value = wave_function(k_example, r_example)

print(f"\nWave Function Value at position {r_example} m:
 ↪ {wf_value:.4f}")
```

This code defines three functions:

- `density_of_states` calculates the density of states based on energy, volume, and electron mass.
- `fermi_dirac_distribution` computes the Fermi-Dirac distribution value for a given energy, chemical potential, and temperature.
- `wave_function` calculates the wave function for free electrons based on the wave vector and position.

The provided example calculates the density of states for a range of energy levels, evaluates the Fermi-Dirac distribution at a specific energy, and computes the wave function at a given position and wave vector, printing the results for analysis.

Chapter 5

Energy Bands

In this chapter, we delve into the fascinating realm of energy bands in crystals and explore their significance in understanding the behavior of electrons in solid materials. The formation of energy bands plays a crucial role in determining the electronic properties and transport phenomena exhibited by various substances. We will investigate the underlying mathematical framework and concepts associated with the formation of energy bands.

Band Theory of Solids

The band theory of solids provides a theoretical framework for understanding the electronic structure of crystalline materials. It postulates that the electronic energy levels in solids form bands, consisting of a large number of closely spaced energy levels. These bands are separated by energy gaps where electron states are not permitted.

1 Tight-Binding Approximation

The tight-binding approximation is a widely adopted approach for modeling energy bands in solids. Within this framework, the electronic wave function is expressed as a linear combination of atomic orbitals, taking into account the overlap between neighboring atoms in the crystal lattice. Mathematically, this can be formulated using the Bloch's theorem as:

$$\psi_{n,\mathbf{k}}(\mathbf{r}) = \sum_j c_{n,\mathbf{k}}(j) e^{i\mathbf{k}\cdot\mathbf{R}_j} \phi(\mathbf{r} - \mathbf{R}_j)$$

Here, $\psi_{n,\mathbf{k}}(\mathbf{r})$ represents the wave function of an electron at energy level $E_{n,\mathbf{k}}$ and crystal momentum $\mathbf{k}$. The coefficients $c_{n,\mathbf{k}}(j)$ account for the contributions from neighboring atomic orbitals $\phi(\mathbf{r} - \mathbf{R}_j)$ at lattice sites $\mathbf{R}_j$. By solving the Schrödinger equation, one can obtain the dispersion relation that describes the energy-momentum relationship within the crystal lattice.

2 Classification of Energy Bands

Energy bands in solids can be classified into three distinct types based on their occupation by electrons: valence bands, conduction bands, and band gaps.

Valence Bands

Valence bands refer to the energy bands that are occupied by valence electrons, which are the outermost electrons of individual atoms. These electrons are tightly bound to the atomic nucleus and are involved in chemical bonding. The valence bands are energetically lower and are separated from the conduction bands by a band gap.

Conduction Bands

Conduction bands, on the other hand, are energy bands that lie above the valence bands and remain unoccupied or partially occupied by electrons at absolute zero temperature. Electrons in the conduction bands are relatively free to move within the crystal and contribute to the electrical conductivity of the material. The energy separation between the conduction bands and valence bands determines the ease of electron movement and, hence, the material's electrical properties.

Band Gaps

The band gap, also known as the energy gap, is the energy range between the valence and conduction bands where no electron states exist. It represents an energy barrier that electrons must overcome to transition from the valence band to the conduction band. Materials can be categorized as either conductors, insulators, or semiconductors based on the width and nature of their band gaps. Conductors have overlapping or partially filled valence and conduction bands, while insulators possess a large band gap, and semiconductors exhibit a small to moderate band gap.

Significance of Energy Bands

The formation and properties of energy bands in crystals have profound implications for the behavior and characteristics of solid materials. Here, we explore the significance of energy bands from a mathematical perspective.

1 Electron Mobility

The existence of energy bands influences the mobility of electrons in a material. In conductors, where the valence and conduction bands overlap, electrons can easily move through the material, resulting in high electrical conductivity. In insulators, the large band gap prevents electron motion, leading to poor electrical conductivity. Semiconductors, with their moderate-sized band gaps, enable controlled electron movement through the application of external stimuli such as heat or electric fields.

2 Electronic and Optical Properties

The energy bands determine various electronic and optical properties of solids. For instance, the electronic band structure affects the material's optical absorption and emission properties. Transitions of electrons between energy bands give rise to phenomena such as absorption of specific wavelengths of light, reflection, and transmission. By analyzing the electronic band structure, one can predict the behavior of materials under different light conditions.

3 Thermal Conductivity

The formation of energy bands also contributes to the understanding of a material's thermal conductivity. The thermal conductivity is influenced by the presence of energy bands and the magnitude of the band gap. In general, materials with wider band gaps tend to exhibit lower thermal conductivity, as thermal energy transfer is hindered by the absence of readily available energy states for electron excitation. Conversely, materials with narrower band gaps or overlapping bands often demonstrate higher thermal conductivity.

Conclusion

Understanding the formation of energy bands in crystals is essential for comprehending the behavior and properties of solid materials. The band theory provides a mathematical framework to analyze the electronic structure and transport phenomena in various substances. Valence bands, conduction bands, and band gaps play critical roles in determining the electrical conductivity, optical properties, and thermal behavior of materials. By exploring the mathematical foundation and significance of energy bands, scientists and engineers can optimize materials for specific applications and discover novel physical phenomena. The next chapter will delve into the derivation and implications of Bloch's theorem, further enhancing our understanding of electron behavior in periodic potentials.

Python Code Snippet

Below is a Python code snippet that calculates the dispersion relation for electrons in a crystal lattice using the tight-binding approximation, as well as estimating the effective mass of electrons in a semiconductor material.

```python
import numpy as np
import matplotlib.pyplot as plt

def calculate_dispersion_relation(t, k_points):
    '''
    Calculate the dispersion relation for a one-dimensional
        tight-binding model.
    :param t: Hopping parameter (in energy units).
    :param k_points: Array of wave vectors.
    :return: Array of energy values corresponding to the k_points.
    '''
    return -2 * t * np.cos(k_points)

def effective_mass(E, k, hbar):
    '''
    Calculate the effective mass of electrons in a semiconductor.
    :param E: Energy value (in energy units).
```

```python
    :param k: Wave vector (in reciprocal space).
    :param hbar: Reduced Planck's constant.
    :return: Effective mass (in mass units).
    """
    # Numerical second derivative to approximate the second
    ↪ derivative of energy with respect to k
    delta_k = 1e-5  # Small change in k
    d2E_dk2 = (calculate_dispersion_relation(t, k + delta_k) -
                2 * calculate_dispersion_relation(t, k) +
                calculate_dispersion_relation(t, k - delta_k)) /
            ↪ (delta_k ** 2)

    return hbar**2 / d2E_dk2

# Constants
t = 1.0  # Hopping parameter
hbar = 1.055e-34  # Reduced Planck's constant (in J*s)
k_points = np.linspace(-np.pi, np.pi, 100)  # Wave vectors

# Calculate dispersion relation
energy_values = calculate_dispersion_relation(t, k_points)

# Plotting the dispersion relation
plt.figure(figsize=(8, 5))
plt.plot(k_points, energy_values, label="Dispersion Relation")
plt.title("Dispersion Relation in a Tight-Binding Model")
plt.xlabel("Wave Vector (k)")
plt.ylabel("Energy (E)")
plt.axhline(0, color='gray', linewidth=0.5, linestyle='--')
plt.axvline(0, color='gray', linewidth=0.5, linestyle='--')
plt.legend()
plt.grid()
plt.show()

# Calculate effective mass for a specific k point
k_value = 0.0  # Example k value
effective_mass_value = effective_mass(0, k_value, hbar)

# Output effective mass
print("Effective Mass of Electron at k =", k_value, "is",
    ↪ effective_mass_value, "kg")
```

This code defines two functions:

- `calculate_dispersion_relation` computes the energy values for a given array of wave vectors using the tight-binding approximation.
- `effective_mass` calculates the effective mass of electrons in a semiconductor material by numerically estimating the second derivative of the energy with respect to the wave vector.

The provided example plots the dispersion relation for a one-dimensional tight-binding model and calculates the effective mass of an electron at a specified wave vector, then prints the result.

Chapter 6

Bloch's Theorem

In this chapter, we will delve into the derivation and implications of Bloch's theorem on the behavior of electrons in periodic potentials. Bloch's theorem is a fundamental result in solid-state physics, providing valuable insight into the electronic structure of crystalline materials. We will begin by presenting the mathematical derivation of Bloch's theorem, followed by its key implications for electron behavior in periodic potentials.

Derivation of Bloch's Theorem

Bloch's theorem establishes a connection between the electronic wave functions in a crystalline solid and the properties of its underlying periodic lattice. The mathematical derivation of Bloch's theorem relies on the concept of translational symmetry in the crystal lattice.

Consider a one-dimensional crystal with a periodic potential defined by a lattice spacing a. The Schrödinger equation governing the behavior of an electron in this periodic potential can be written as:

$$\hat{H}\psi(r) = E\psi(r)$$

where $\hat{H}$ is the Hamiltonian operator, $\psi(r)$ is the wave function, E is the energy eigenvalue, and r represents the spatial variable.

In order to explore the solutions of this equation, we assume that the wave function has a specific form given by:

$$\psi(r) = u(r)e^{ik\cdot r}$$

where k is the crystal momentum, and $u(r)$ is a periodic function with the same periodicity as the crystal lattice. This periodic function can be expressed as:

$$u(r + a) = u(r)$$

By substituting this wave function into the Schrödinger equation, we arrive at:

$$\hat{H}u(r)e^{ik\cdot r} = Eu(r)e^{ik\cdot r}$$

Since $e^{ik\cdot r}$ has a periodicity matching that of the lattice, it can be factored out:

$$\hat{H}u(r) = (E - k^2)u(r)$$

This equation reveals that the function $u(r)$ satisfies an eigenvalue equation with an effective potential given by $(E-k^2)$. Hence, the problem of solving the Schrödinger equation in a periodic potential has been reduced to finding the eigenstates of this effective potential.

Implications for Electrons in Periodic Potentials

Bloch's theorem provides valuable insights into the behavior of electrons in periodic potentials and has significant implications for the electronic structure of crystalline materials.

1 Band Structure

The solutions of the effective potential equation $(E-k^2)$ give rise to energy levels known as bands. The energy bands in a periodic crystal are formed by the eigenstates of the effective potential, which exhibit a certain periodicity within the crystalline lattice. These energy bands and their corresponding band gaps determine the electronic properties of the material, such as electrical conductivity and optical behavior.

2 Energy-Momentum Relationship

Bloch's theorem establishes a clear relationship between the energy eigenvalues and the crystal momentum. The energy-momentum relationship of electrons in a crystal is described by a dispersion relation. This relation shows how the energy of electrons varies with their momentum and provides valuable information about the electronic properties of the material, such as its effective mass and group velocity.

3 Density of States

By considering the periodicity of the wave function, Bloch's theorem provides insights into the density of states (DOS) in a crystal. The DOS represents the number of energy states available to electrons at each energy level in the material. Understanding the DOS is crucial for analyzing various phenomena in crystalline solids, including electrical conductivity, thermal properties, and electronic transitions.

4 Band Gaps and Electronic Conductivity

The existence of band gaps in the energy spectrum of a crystal is a direct consequence of its periodicity. These band gaps play a crucial role in determining the electrical conductivity of materials. In insulators, a large band gap prevents the movement of electrons, resulting in low electrical conductivity. In contrast, conductors possess overlapping energy bands, allowing electrons to move freely and facilitating high electrical conductivity. Semiconductors, with their moderate-sized band gaps, exhibit intermediate conductivity properties and can be easily modulated through external influences.

Conclusion

Bloch's theorem, derived from the translational symmetry of periodic lattices, provides valuable insights into the behavior of electrons in crystalline materials. It establishes a connection between the wave functions, energy eigenvalues, and crystal momentum, enabling a comprehensive understanding of the electronic structure and properties of periodic solids. The implications of Bloch's theorem, including the formation of energy bands, the energy-momentum relationship, density of states, and band gaps, contribute to our comprehension of a wide range of phenomena, from electrical conductivity to optical behavior."'latex

Python Code Snippet

Below is a Python code snippet that implements key equations and algorithms related to Bloch's theorem. This code simulates the behavior of electrons in a one-dimensional periodic potential and computes the effective potential, energy bands, and density of states.

```python
import numpy as np
import matplotlib.pyplot as plt

def bloch_wavefunction(x, k, a):
    '''
    Calculate the Bloch wavefunction for a one-dimensional crystal.

    :param x: Position array in meters.
    :param k: Crystal momentum vector.
    :param a: Lattice spacing of the crystal.
    :return: Bloch wavefunction values.
    '''
    return np.cos(k * x) * np.exp(1j * k * x)

def effective_potential(E, k):
    '''
    Compute the effective potential based on energy eigenvalue.

    :param E: Energy eigenvalue.
    :param k: Crystal momentum vector.
    :return: Effective potential.
    '''
    # Assuming an otherwise free electron and quadratic dependence
    return E - (h_bar**2 * k**2) / (2 * me)

def density_of_states(E):
    '''
    Calculate the Density of States (DOS) at a given energy E.

    :param E: Energy level.
    :return: Density of states value.
    '''
    # For a 1D free electron gas
    return (1 / np.pi) * (2 * me / h_bar**2)**(1/2) * \
        np.abs(E)**(1/2)

# Constants
h_bar = 1.0545718e-34    # Reduced Planck constant (J.s)
```

```python
me = 9.10938356e-31      # Electron rest mass (kg)
a = 1e-9                 # Lattice spacing (meters)
k_values = np.linspace(-1e10, 1e10, 100)  # Momentum range

# Energy and effective potential calculations
E_values = np.array([effective_potential(E, k) for k in k_values])
DOS_values = np.array([density_of_states(E) for E in E_values])

# Plotting the results
plt.figure(figsize=(12, 6))

# Energy vs Momentum
plt.subplot(1, 2, 1)
plt.plot(k_values, E_values)
plt.title('Effective Potential vs Crystal Momentum')
plt.xlabel('Crystal Momentum (kg m/s)')
plt.ylabel('Effective Potential (J)')
plt.grid()

# Density of States
plt.subplot(1, 2, 2)
plt.plot(E_values, DOS_values)
plt.title('Density of States vs Energy')
plt.xlabel('Energy (J)')
plt.ylabel('Density of States (1/J)')
plt.grid()

plt.tight_layout()
plt.show()
```

This code defines three functions:

- `bloch_wavefunction` computes the Bloch wavefunction for given positions and crystal momenta.
- `effective_potential` calculates the effective potential based on the energy eigenvalue and crystal momentum.
- `density_of_states` computes the density of states for a given energy level.

The provided example simulates the effective potential and density of states for electrons within a one-dimensional periodic potential and visualizes the results using plots. Make sure to set appropriate units for the parameters based on your application. "'

Chapter 7

Brillouin Zones

The concept of Brillouin zones plays a crucial role in the band theory of solids, providing valuable insights into the electronic structure of crystalline materials. Brillouin zones are a powerful tool for analyzing the behavior of electrons in periodic potentials and understanding various phenomena, such as energy bands and electronic transport properties. In this chapter, we will explore the concept and importance of Brillouin zones in the band theory of solids.

Periodic Potentials and Reciprocal Space

To understand Brillouin zones, we must first establish a connection between periodic potentials and reciprocal space. In solids, the wave function of an electron exhibits a periodicity matching that of the crystal lattice. This periodicity gives rise to a reciprocal lattice in reciprocal space, which serves as a mathematical representation of the translational symmetry in the crystal lattice.

The reciprocal lattice is defined as the set of all wave vectors $\mathbf{K}$ that satisfy the equation:

$$e^{i\mathbf{K}\cdot\mathbf{R}} = 1$$

where $\mathbf{R}$ denotes the position vector of any lattice point in the real lattice. The reciprocal lattice vectors $\mathbf{K}$ are linearly related to the lattice vectors $\mathbf{a}$ of the real lattice through a set of coefficients called the reciprocal lattice basis vectors $\mathbf{b}$:

$$\mathbf{K} = h_1\mathbf{b_1} + h_2\mathbf{b_2} + h_3\mathbf{b_3}$$

where h_1, h_2, and h_3 are integers.

Brillouin Zones and the First Brillouin Zone

Brillouin zones are the regions of reciprocal space that correspond to unique physical properties of a crystal lattice. The concept of Brillouin zones is based on the observation that the periodicity of the lattice in real space is directly related to the periodicity of the reciprocal lattice in reciprocal space.

The first Brillouin zone (FBZ) is the Wigner-Seitz cell of the reciprocal lattice, which is constructed by connecting the midpoints of the reciprocal lattice vectors to form a polyhedral region. The FBZ is the smallest polyhedron in reciprocal space that fully encloses a lattice point at the origin. It contains all the wave vectors that can be added to a reciprocal lattice vector without reaching another reciprocal lattice point.

The FBZ plays a fundamental role in the band structure of solids. The electronic behavior and energy band properties of a crystal can be fully described by considering the electronic states within the FBZ. The FBZ also determines the periodicity of energy bands in reciprocal space, providing valuable information about the energy-momentum relations, density of states, and transport properties of electrons in crystalline materials.

Higher Brillouin Zones

In addition to the first Brillouin zone, higher Brillouin zones exist in reciprocal space. The second, third, and subsequent Brillouin zones are constructed by connecting the midpoints of the reciprocal lattice vectors with their nearest neighbors. Each Brillouin zone represents a different set of wave vectors that satisfy the periodicity condition of the crystal lattice.

The existence of higher Brillouin zones reflects the multiple solutions to the wave equation in a periodic potential. While the FBZ contains the essential information about the band structure, higher Brillouin zones provide further insight into the behavior of electrons in the crystal lattice. The higher Brillouin zones are critical for understanding phenomena such as energy dispersion, wave interference, and electronic transitions in periodic solids.

Importance in Band Theory

The concept of Brillouin zones is of paramount importance in the band theory of solids. The band structure, which describes the energy levels and allowed electronic states of a crystal, can be fully characterized within the FBZ. The FBZ determines the energy-momentum relationship, density of states, and transport properties of electrons in the crystal lattice. It provides a powerful framework for understanding and predicting various electronic phenomena, such as electrical conductivity, thermal properties, and optical behavior.

Furthermore, Brillouin zones help explain the origin of energy gaps, energy dispersion, and wave interference effects in crystalline materials. By considering the electronic states within the FBZ and higher Brillouin zones, one can gain insights into the nature of energy bands, the localization of electronic wave functions, and the occurrence of band gaps or band overlaps.

The knowledge of Brillouin zones also enables the design and analysis of novel electronic devices based on the band structure of solids. By controlling the Brillouin zone boundaries, scientists and engineers can tailor the electronic properties of materials, manipulate the density of states, and enhance the performance of electronic and optoelectronic devices.

Thus, Brillouin zones offer a powerful mathematical framework for studying the electronic structure, energy bands, and transport properties of crystalline solids. Their importance in the band theory of solids cannot be overstated, making them an indispensable tool for researchers and engineers working in the field of condensed matter physics.

Conclusion

In this chapter, we have explored the concept and importance of Brillouin zones in the band theory of solids. Brillouin zones allow us to analyze the behavior of electrons in periodic potentials and understand various phenomena, such as energy bands and electronic transport properties. The first Brillouin zone serves as a

fundamental region in reciprocal space, containing all the essential information about the band structure of materials. Higher Brillouin zones provide further insights into the electronic behavior within the crystal lattice. The knowledge of Brillouin zones is crucial for understanding the electronic structure of solids, designing electronic devices, and predicting material properties. By leveraging the concept and mathematical framework of Brillouin zones, researchers and engineers continue to uncover new possibilities for advancing the field of condensed matter physics.Sure! Here is the section containing a comprehensive Python code snippet based on the important equations and algorithms mentioned in the chapter on Brillouin zones, properly wrapped in LaTeX using the minted package:
"'latex

Python Code Snippet

Below is a Python code snippet that calculates the Brillouin Zone boundaries, generates a reciprocal lattice vector, and visualizes the first Brillouin zone along with higher Brillouin zones for a given crystal lattice.

```python
import numpy as np
import matplotlib.pyplot as plt

def generate_reciprocal_lattice_vectors(a1, a2):
    '''
    Generate reciprocal lattice vectors given the primitive lattice
      vectors in real space.
    :param a1: First lattice vector in real space (2D, numpy array).
    :param a2: Second lattice vector in real space (2D, numpy
      array).
    :return: Reciprocal lattice vectors (numpy array).
    '''
    volume = np.dot(a1, np.cross(a2, np.array([0,0,1])))  # Volume
      of the unit cell
    b1 = (2 * np.pi / volume) * np.array([0, 0, 1])  # Reciprocal
      vector b1
    b2 = (2 * np.pi / volume) * np.array([0, 0, 1])  # Reciprocal
      vector b2
    b1[0] = (2 * np.pi) * (a2[1] / volume)
    b1[1] = -(2 * np.pi) * (a1[0] / volume)
    
    b2[0] = -(2 * np.pi) * (a1[1] / volume)
    b2[1] = (2 * np.pi) * (a2[0] / volume)
```

```python
    return np.array([b1, b2])

def first_brillouin_zone(a1, a2):
    '''
    Compute and visualize the first Brillouin zone from the given
    ↪ lattice vectors.
    :param a1: First lattice vector in real space (2D, numpy array).
    :param a2: Second lattice vector in real space (2D, numpy
    ↪ array).
    '''
    brillouin_boundary = []
    b1, b2 = generate_reciprocal_lattice_vectors(a1, a2)

    # Generating points for the first Brillouin zone
    for i in np.linspace(-1.5, 1.5, 100):
        for j in np.linspace(-1.5, 1.5, 100):
            k_point = i * b1 + j * b2
            if np.linalg.norm(k_point) < np.pi:   # Within the first
            ↪ Brillouin zone
                brillouin_boundary.append(k_point)

    brillouin_boundary = np.array(brillouin_boundary)

    plt.figure(figsize=(8, 8))
    plt.scatter(brillouin_boundary[:, 0], brillouin_boundary[:, 1],
    ↪ color='b', s=1)
    plt.xlim(-np.pi, np.pi)
    plt.ylim(-np.pi, np.pi)
    plt.axhline(0, color='black', lw=0.5)
    plt.axvline(0, color='black', lw=0.5)
    plt.title('First Brillouin Zone')
    plt.xlabel('k_x')
    plt.ylabel('k_y')
    plt.grid()
    plt.show()

# Inputs for the lattice vectors
a1 = np.array([1.0, 0.0])   # First lattice vector
a2 = np.array([0.0, 1.0])   # Second lattice vector

# Visualize the first Brillouin zone
first_brillouin_zone(a1, a2)
```

This code defines two functions:

- `generate_reciprocal_lattice_vectors` computes the reciprocal lattice vectors using the provided primitive lattice vectors in real space.
- `first_brillouin_zone` visualizes the first Brillouin zone by computing the k-points within the zone and plotting them.

The provided example uses the primitive lattice vectors in real space to generate and visualize the first Brillouin zone corresponding to a given crystal lattice. "'

This LaTeX code provides a clear and structured Python code snippet that calculates and visualizes the first Brillouin zone, making it easy for readers to follow along and understand the applicable concepts.

Chapter 8

Effective Mass Theorem

In this chapter, we delve into the concept of effective mass in semiconductors, a fundamental quantity that describes the behavior of charge carriers in the presence of a crystal lattice. The effective mass provides crucial insights into the electrical conductivity and transport properties of semiconductors, enabling the analysis and design of semiconductor devices. We explore the theory behind the effective mass and its applications in the study of semiconductor physics.

Understanding Effective Mass

In the study of semiconductors, the interaction between charge carriers (electrons and holes) and the crystal lattice plays a significant role in determining their behavior. To simplify this complex interaction, a concept known as the effective mass is introduced. The effective mass reflects how charge carriers behave as though they have a mass different from their free-electron mass due to the influence of the crystal lattice.

Mathematically, the effective mass is defined by the proportionality between the acceleration of a charge carrier and the applied force:

$$\mathbf{F} = m_{\text{eff}} \cdot \mathbf{a}$$

where **F** is the force acting on the charge carrier, **a** is the resulting acceleration, and m_eff represents the effective mass. In general, the effective mass can vary depending on the energy and direction of the charge carrier.

Effective Mass Approximation

To facilitate the computation of the effective mass, an approximation known as the effective mass approximation is often employed. This approximation assumes that the energy dispersion relation near the band extrema in the crystal can be approximated by a parabolic function. Under this assumption, one can define the effective mass as the inverse of the curvature of the energy band:

$$m_\text{eff} = \left(\frac{1}{\hbar^2} \frac{d^2 E}{dk^2} \right)^{-1}$$

where E represents the energy of the charge carrier, k is the wave vector, and $\hbar$ is the reduced Planck's constant. This definition implies that a large effective mass corresponds to a small curvature of the energy band, indicating less dispersion and a slower mobility for the charge carrier.

Anisotropic Effective Mass

In crystalline materials, the effective mass can exhibit anisotropic behavior, meaning that it can vary depending on the crystallographic direction. In these cases, the effective mass is expressed as a tensor, denoted as m^*_{ij}. The tensor provides information about the effective mass along different crystallographic directions.

Typically, the tensor is diagonal, meaning it has non-zero elements only on the main diagonal. The diagonal elements, m^*_{ii}, represent the effective masses along the respective crystallographic axes. The off-diagonal elements, m^*_{ij} with $i \neq j$, are zero in most cases due to the symmetry properties of the crystal.

Characteristics of Effective Mass

The effective mass possesses several important properties that make it a valuable quantity for the study of semiconductors. First, it determines the mobility of charge carriers, which quantifies how easily they move in response to an applied electric field. The effective mass serves as a key parameter in the calculation of carrier mobility, providing vital information for the design and optimization of semiconductor devices.

Second, the effective mass affects the density of states in the energy bands of semiconductors. The density of states determines the availability of energy states to accommodate charge carriers. By manipulating the effective mass, one can control the density of states and influence the electrical and optical properties of semiconductor materials.

Lastly, the effective mass plays a crucial role in the analysis of scattering mechanisms in semiconductors. Scattering events, such as those from impurities, lattice defects, or phonons, affect the mobility of charge carriers. The effective mass enters into the scattering rate calculations, allowing researchers to understand and optimize the material's performance.

Applications of Effective Mass

The effective mass concept finds extensive applications in the field of semiconductor physics and device engineering. One prominent application is the modeling and characterization of semiconductors, where the effective mass plays a central role in simulating charge carrier behavior and optimizing device performance. It allows researchers to predict and understand phenomena such as carrier dynamics, carrier transport, and energy band structures.

The effective mass also aids in the design and analysis of semiconductor devices, including transistors, solar cells, and integrated circuits. By precisely controlling the effective mass of charge carriers, engineers can tailor the device characteristics and improve their operational efficiency.

Furthermore, the effective mass proves valuable in materials engineering, allowing for the identification and optimization of novel semiconductor materials with desired electrical, optical, and mechanical properties. Researchers can assess material properties, such as effective mass anisotropy, to select suitable materials for specific applications.

Conclusion

In this chapter, we have explored the concept of effective mass in semiconductors and its significance in the study of semiconductor physics. The effective mass approximation provides a useful tool for understanding the behavior of charge carriers in crystalline materials, enabling the analysis and design of semiconductor devices. With its various applications in simulating charge carrier behavior, optimizing device performance, and materials engineering, the effective mass continues to be a fundamental concept in the field of semiconductor physics.

Python Code Snippet

Below is a Python code snippet that calculates the effective mass, mobility, and scattering parameters for charge carriers in semiconductors based on the formulas derived in this chapter.

```python
import numpy as np

def calculate_effective_mass(energy, hbar):
    '''
    Calculate the effective mass using the second derivative of
      energy with respect to wave vector.
    :param energy: Energy of the charge carrier in joules.
    :param hbar: Reduced Planck's constant in joules per second.
    :return: Effective mass in kilograms.
    '''
    k_values = np.linspace(-1e10, 1e10, num=1000)  # Wave vector
      range
    energies = hbar * k_values**2 / (2 * calculate_curvature(energy,
      k_values))  # Energy for effective mass calculation
    m_eff = hbar**2 / (np.gradient(np.gradient(energies), k_values))
      # Effective mass calculation
    return m_eff.mean()  # Return the average effective mass


def calculate_curvature(energy, k_values):
    '''
    Calculate curvature as the second derivative of energy with
      respect to wave vector.
    :param energy: Energy of the charge carrier.
    :param k_values: Wave vector values.
    :return: Curvature value.
    '''
    return np.gradient(np.gradient(energy, k_values), k_values)
```

```python
def calculate_mobility(effective_mass, charge, temperature):
    '''
    Calculate the mobility of charge carriers in a semiconductor.
    :param effective_mass: Effective mass of charge carriers in kg.
    :param charge: Charge of the carrier in coulombs.
    :param temperature: Temperature in Kelvin.
    :return: Mobility in m²/(V·s).
    '''
    k_B = 1.38064852e-23  # Boltzmann constant in J/K
    mobility = (charge * temperature) / (k_B * effective_mass)  #
    ↪ Mobility calculation
    return mobility


def scattering_rate(effective_mass, density, temperature):
    '''
    Calculate the scattering rate of charge carriers.
    :param effective_mass: Effective mass of the carriers in kg.
    :param density: Carrier density in m³.
    :param temperature: Temperature in Kelvin.
    :return: Scattering rate in 1/s.
    '''
    return (density * temperature) / effective_mass  # Scattering
    ↪ rate calculation


# Constants and sample inputs
hbar = 1.0545718e-34  # Reduced Planck's constant in J·s
charge = 1.602176634e-19  # Charge of an electron in C
temperature = 300  # Temperature in K (room temperature)
density = 1e25  # Carrier density in m³

# Calculate effective mass
energy = 1e-19  # Example energy in joules
effective_mass = calculate_effective_mass(energy, hbar)

# Calculate mobility
mobility = calculate_mobility(effective_mass, charge, temperature)

# Calculate scattering rate
sc_rate = scattering_rate(effective_mass, density, temperature)

# Output results
print("Effective Mass:", effective_mass, "kg")
print("Mobility:", mobility, "m²/(V·s)")
print("Scattering Rate:", sc_rate, "1/s")
```

This code defines four functions:

- `calculate_effective_mass` computes the effective mass of charge

carriers based on energy and the reduced Planck's constant.
- `calculate_mobility` determines the mobility of charge carriers using the effective mass, charge, and temperature.
- `scattering_rate` calculates the scattering rate of charge carriers given the effective mass, carrier density, and temperature.
- `calculate_curvature` provides the second derivative of energy for effective mass calculation.

The provided example calculates the effective mass, mobility, and scattering rate for charge carriers in semiconductors, then prints the results.

Chapter 9

Density of States

In this chapter, we will delve into the derivation of the density of states for electrons and holes in semiconductors. The density of states is a fundamental concept in the study of semiconductor physics, providing valuable insights into the available energy states for charge carriers. By understanding the density of states, we can analyze the electrical and optical properties of semiconductors, and gain a deeper understanding of their behavior.

Energy Bands and Energy States

In semiconductors, the energy of a charge carrier is distributed among different energy levels or states. These energy states are formed due to the interaction of electrons with the crystal lattice. Electrons can occupy these states, giving rise to energy bands.

The density of states is defined as the number of energy states per unit energy range, per unit volume. It provides a measure of the available energy states for charge carriers within a given energy range. To derive the density of states, we need to consider the energy dispersion relation and the distribution of energy states in the semiconductor.

1 Energy Dispersion Relation

The energy dispersion relation relates the energy of a charge carrier to its wave vector. In semiconductors, the energy dispersion relation can be approximated as parabolic near the band extrema. For electrons in the conduction band, the energy dispersion relation takes the form:

$$E_c(\mathbf{k}) = E_{c0} + \frac{\hbar^2(\mathbf{k} - \mathbf{k}_{c0})^2}{2m_c^*}$$

where $E_c(\mathbf{k})$ is the energy of an electron in the conduction band, $\mathbf{k}$ is the wave vector, E_{c0} is the energy at the band minimum, $\mathbf{k}_{c0}$ is the wave vector at the band minimum, and m_c^* is the effective mass of the electron.

Similarly, for holes in the valence band, the energy dispersion relation is given by:

$$E_v(\mathbf{k}) = E_{v0} - \frac{\hbar^2(\mathbf{k} - \mathbf{k}_{v0})^2}{2m_v^*}$$

where $E_v(\mathbf{k})$ is the energy of a hole in the valence band, $\mathbf{k}$ is the wave vector, E_{v0} is the energy at the band maximum, $\mathbf{k}_{v0}$ is the wave vector at the band maximum, and m_v^* is the effective mass of the hole.

2 Distribution of Energy States

The distribution of energy states in a semiconductor depends on the available energy levels for charge carriers. To calculate the density of states, we need to consider the number of energy states per unit energy range, per unit volume. This can be determined using the relationship between the wave vector and energy.

For electrons, the number of energy states in a small energy range dE per unit volume is given by:

$$g_c(E)dE = \frac{V \cdot d\mathbf{k}}{\left(\frac{d^2 E_c}{d\mathbf{k}^2}\right)^{-1}}$$

where $g_c(E)$ is the density of states for electrons, V is the volume of the semiconductor, $d\mathbf{k}$ is the infinitesimal change in wave vector, and $\frac{d^2 E_c}{d\mathbf{k}^2}$ represents the second derivative of the energy dispersion relation.

Similarly, for holes, the density of states is given by:

$$g_v(E)dE = \frac{V \cdot d\mathbf{k}}{\left(\frac{d^2 E_v}{d\mathbf{k}^2}\right)^{-1}}$$

where $g_v(E)$ is the density of states for holes.

Derivation of Density of States

To derive the density of states, we need to express the wave vector in terms of energy using the energy dispersion relation. We will consider a one-dimensional case for simplicity.

For electrons, using the energy dispersion relation $E_c(k)$, we can express the wave vector k in terms of energy E as:

$$k = \sqrt{\frac{2m_c^*}{\hbar^2}(E - E_{c0})}$$

Substituting this expression for k in the equation for $g_c(E)$, we obtain:

$$g_c(E)dE = \frac{V \cdot dk}{d^2 E_c / dk^2}$$

To evaluate dk, we differentiate the expression for k with respect to E:

$$dk = \frac{m_c^*}{\hbar^2 \sqrt{\frac{2m_c^*}{\hbar^2}(E - E_{c0})}} dE$$

Expanding $\frac{d^2 E_c}{dk^2}$, and substituting the expression for dk, we obtain:

$$g_c(E)dE = \frac{V \cdot \sqrt{\frac{2m_c^*}{\hbar^2}}(E - E_{c0})^{-\frac{3}{2}}}{\frac{d^2 E_c}{dk^2}} dE$$

A similar derivation can be done for holes, yielding the expression for $g_v(E)$:

$$g_v(E)dE = \frac{V \cdot \sqrt{\frac{2m_v^*}{\hbar^2}}(E_{v0} - E)^{-\frac{3}{2}}}{\frac{d^2 E_v}{dk^2}} dE$$

Properties of the Density of States

The density of states exhibits certain properties that are important for understanding the behavior of charge carriers in semiconductors.

Firstly, the density of states diverges as the energy approaches the band edges. This divergence occurs due to the Van Hove singularities, which arise from the vanishing of the second derivative of the energy dispersion relation at the band extrema.

Secondly, the density of states varies with the effective mass of the charge carriers. A smaller effective mass results in a larger density of states, leading to increased carrier concentration and higher conductivity.

Lastly, the density of states affects various physical phenomena such as carrier concentration, carrier transport, and optical transitions in semiconductors. By manipulating the density of states through material engineering or external factors like doping, researchers can control and optimize the performance of semiconductor devices.

Conclusion

In this chapter, we have derived the density of states for electrons and holes in semiconductors. We expressed the density of states in terms of the energy dispersion relation and the distribution of energy states. The density of states provides crucial information about the available energy states for charge carriers and plays a significant role in understanding their behavior in semiconductors. The properties of the density of states allow researchers to manipulate and optimize charge carrier concentrations, carrier transport properties, and optical phenomena in semiconductor materials.

Python Code Snippet

Below is a Python code snippet that calculates the density of states for electrons and holes in semiconductors based on the concepts discussed in this chapter.

```
import numpy as np

def density_of_states_electrons(E, m_c_star, E_c0):
    '''
    Calculate the density of states for electrons in the conduction
     band.
    :param E: Energy level in joules.
    :param m_c_star: Effective mass of the electron in kg.
    :param E_c0: Energy at the conduction band minimum in joules.
```

```
    :return: Density of states for electrons in states/m^3/J.
    '''
    if E < E_c0:
        return 0  # No states below the conduction band minimum
    k = np.sqrt(2 * m_c_star * (E - E_c0)) / np.hbar
    dk_dE = (m_c_star / (np.hbar ** 2 * np.sqrt(2 * m_c_star * (E -
    ↪     E_c0))))  # Derivative of k w.r.t E
    dE_dk = 1 / (m_c_star / (np.hbar ** 2 * k))  # Inverse of 2nd
    ↪     derivative of E w.r.t k
    g_c = (2 * (2 / np.pi**2) * (k**2)) * dk_dE * dE_dk
    return g_c

def density_of_states_holes(E, m_v_star, E_v0):
    '''
    Calculate the density of states for holes in the valence band.
    :param E: Energy level in joules.
    :param m_v_star: Effective mass of the hole in kg.
    :param E_v0: Energy at the valence band maximum in joules.
    :return: Density of states for holes in states/m^3/J.
    '''
    if E > E_v0:
        return 0  # No states above the valence band maximum
    k = np.sqrt(2 * m_v_star * (E_v0 - E)) / np.hbar
    dk_dE = (m_v_star / (np.hbar ** 2 * np.sqrt(2 * m_v_star * (E_v0
    ↪     - E))))  # Derivative of k w.r.t E
    dE_dk = 1 / (m_v_star / (np.hbar ** 2 * k))  # Inverse of 2nd
    ↪     derivative of E w.r.t k
    g_v = (2 * (2 / np.pi**2) * (k**2)) * dk_dE * dE_dk
    return g_v


# Constants
hbar = 1.0545718e-34  # Reduced Planck's constant in J.s
m_electron = 9.10938356e-31  # Mass of electron in kg
m_hole = 9.10938356e-31 * 0.5  # Assume effective mass of hole is
↪     half of electron's mass

# Inputs for the calculations
E_c0 = 1.1 * 1.602e-19  # Band gap energy for intrinsic silicon in
↪     joules (1.1 eV)
E_v0 = 0  # Assume the maximum energy of the valence band is 0 J
energy_levels = np.linspace(E_v0 - 0.1 * 1.602e-19, E_c0 + 0.1 *
↪     1.602e-19, 100)  # Energy levels around the band edges

# Calculate density of states for electrons and holes
density_states_electrons =
↪     density_of_states_electrons(energy_levels, m_electron, E_c0)
density_states_holes = density_of_states_holes(energy_levels,
↪     m_hole, E_v0)

# Output results
import matplotlib.pyplot as plt
```

```
plt.figure(figsize=(10, 5))
plt.plot(energy_levels / 1.602e-19, density_states_electrons,
    label='Density of States (Electrons)', color='blue')
plt.plot(energy_levels / 1.602e-19, density_states_holes,
    label='Density of States (Holes)', color='red')
plt.axvline(E_c0 / 1.602e-19, color='blue', linestyle='--',
    label='Conduction Band Edge')
plt.axvline(E_v0 / 1.602e-19, color='red', linestyle='--',
    label='Valence Band Edge')
plt.title('Density of States for Electrons and Holes')
plt.xlabel('Energy (eV)')
plt.ylabel('Density of States (states/m³/J)')
plt.legend()
plt.grid()
plt.show()
```

This code defines two primary functions:

- `density_of_states_electrons` calculates the density of states for electrons in the conduction band based on the energy, effective mass, and conduction band edge.
- `density_of_states_holes` computes the density of states for holes in the valence band using similar parameters.

The code generates a plot showing the density of states for both electrons and holes across a range of energy levels, illustrating their behavior around the band edges, which is critical for semiconductor applications.

Chapter 10

Fermi-Dirac Statistics

In this chapter, we explore the application of Fermi-Dirac statistics to semiconductors, specifically examining its significance in determining the intrinsic carrier concentration. Fermi-Dirac statistics play a crucial role in understanding the behavior of electrons and holes in semiconductors at thermal equilibrium. By applying this statistical framework, we can gain insights into the occupation probabilities of energy states and quantify the number of charge carriers in various energy levels within a semiconductor material.

Fermi-Dirac Distribution Function

The Fermi-Dirac distribution function describes the probability of finding a fermion (a particle with half-integer spin) in a particular energy state at a given temperature. For semiconductors, we are primarily interested in electrons and holes, which are both fermionic particles.

The Fermi-Dirac distribution function for electrons is given by:

$$f_e(E) = \frac{1}{1 + e^{(E-E_{Fe})/(k_B T)}}$$

where $f_e(E)$ is the probability that a given energy state E is occupied by an electron, E_{Fe} is the Fermi energy level for electrons, k_B is Boltzmann's constant, and T is the temperature.

Similarly, for holes, the Fermi-Dirac distribution function is:

$$f_h(E) = \frac{1}{1 + e^{(E_{Fh}-E)/(k_B T)}}$$

where $f_h(E)$ is the probability that a given energy state E is occupied by a hole, and E_{Fh} is the Fermi energy level for holes.

Occupation Probabilities in Semiconductors

In semiconductors at thermal equilibrium, the occupation probabilities of energy states for both electrons and holes follow the Fermi-Dirac distribution function. These probabilities provide crucial information about the number of electrons and holes in different energy levels, which in turn determines the intrinsic carrier concentration.

The intrinsic carrier concentration, denoted by n_i, represents the number of electrons in the conduction band and the number of holes in the valence band per unit volume in an undoped or intrinsic semiconductor. At thermal equilibrium, the occupation probabilities for electrons and holes can be expressed as:

$$f_e(E) = \frac{1}{1 + e^{(E-E_{Fe})/(k_B T)}} \quad \text{and} \quad f_h(E) = \frac{1}{1 + e^{(E_{Fh}-E)/(k_B T)}}$$

Due to the energy bandgap, the Fermi energy level for electrons and holes is different. For an intrinsic semiconductor, where the electron and hole concentrations are equal, the Fermi energy level E_{Fi} can be expressed as the average of the conduction band minimum energy E_c and the valence band maximum energy E_v:

$$E_{Fi} = \frac{1}{2}(E_c + E_v)$$

Using the principles of charge neutrality, the intrinsic carrier concentration n_i can be approximated as:

$$n_i = N_c \cdot N_v \cdot e^{-\frac{E_g}{k_B T}}$$

where N_c is the effective density of states in the conduction band, N_v is the effective density of states in the valence band, and E_g is the energy bandgap.

Importance in Semiconductor Physics

Fermi-Dirac statistics and the intrinsic carrier concentration are of paramount importance in the field of semiconductor physics.

By understanding the occupation probabilities of energy states, researchers can analyze and predict the behavior of electrons and holes in semiconductors at thermal equilibrium. The intrinsic carrier concentration provides a baseline for the number of charge carriers within a material, influencing various physical properties such as electrical conductivity, optical absorption, and energy band diagrams.

Moreover, the Fermi-Dirac distribution function allows for the calculation of other quantities related to charge carrier statistics, such as the Fermi energy level E_F, which represents the energy at which the occupation probability is exactly 0.5. The position of the Fermi energy level relative to the energy bands is crucial in determining the availability of states for charge carriers and plays a significant role in device operation and material design.

In summary, the application of Fermi-Dirac statistics to semiconductors, specifically in determining the intrinsic carrier concentration, is essential in understanding the behavior of electrons and holes at thermal equilibrium. By characterizing the occupation probabilities of energy states, researchers can gain insights into the number of charge carriers and their impact on semiconductor device performance.

Python Code Snippet

Below is a Python code snippet that calculates the intrinsic carrier concentration in a semiconductor using the Fermi-Dirac statistics and other important equations discussed in this chapter.

```python
import numpy as np

def fermi_dirac_distribution(E, E_F, T):
    '''
    Calculate the Fermi-Dirac distribution for electrons.
    :param E: Energy level (Joules).
    :param E_F: Fermi energy level (Joules).
    :param T: Temperature (Kelvins).
    :return: Occupation probability for electrons.
    '''
    k_B = 1.38e-23  # Boltzmann's constant in J/K
    return 1 / (1 + np.exp((E - E_F) / (k_B * T)))

def intrinsic_carrier_concentration(N_c, N_v, E_g, T):
    '''
    Calculate the intrinsic carrier concentration in semiconductors.
```

```
:param N_c: Effective density of states in the conduction band
    (m^-3).
:param N_v: Effective density of states in the valence band
    (m^-3).
:param E_g: Energy bandgap (Joules).
:param T: Temperature (Kelvins).
:return: Intrinsic carrier concentration (m^-3).
'''
k_B = 1.38e-23   # Boltzmann's constant in J/K
return N_c * N_v * np.exp(-E_g / (2 * k_B * T))

# Constants and parameters
E_c = 1.1 * 1.6e-19   # Conduction band minimum energy in Joules (1.1
    eV)
E_v = 0 * 1.6e-19     # Valence band maximum energy in Joules (0 eV)
E_g = E_c - E_v       # Energy bandgap
T = 300               # Temperature in Kelvin

# Effective density of states
N_c = 1e25            # Effective density of states in the conduction
    band (m^-3)
N_v = 6e24            # Effective density of states in the valence
    band (m^-3)

# Calculate Fermi energy level for intrinsic semiconductor
E_F = (E_c + E_v) / 2   # Average energy for intrinsic carrier
    concentration

# Calculation of intrinsic carrier concentration
n_i = intrinsic_carrier_concentration(N_c, N_v, E_g, T)

# Calculate the occupation probabilities
E = np.linspace(E_v, E_c, 100)   # Energy levels between valence and
    conduction bands
f_e = fermi_dirac_distribution(E, E_F, T)   # Occupation
    probabilities for electrons
f_h = fermi_dirac_distribution(E, E_F, T - 0.1)   # For holes,
    assuming the same Fermi level for simplicity

# Output results
print("Intrinsic Carrier Concentration (n_i):", n_i, "m^-3")
print("Fermi Energy Level (E_F):", E_F * 1e19, "eV")   # Convert to
    eV for output
print("Occupancy probabilities for selected energy levels:")
for i in range(0, 100, 20):   # Show every 20th value
    print(f"Energy: {E[i] * 1e19:.2f} eV, Electron occupancy:
        {f_e[i]:.4f}, Hole occupancy: {f_h[i]:.4f}")
```

This code defines two main functions:

- fermi_dirac_distribution computes the occupation probabil-

ity of energy states for electrons using the Fermi-Dirac distribution.
- `intrinsic_carrier_concentration` calculates the intrinsic carrier concentration in a semiconductor based on the effective density of states, energy bandgap, and temperature.

The provided example calculates the intrinsic carrier concentration, the Fermi energy level, and prints the occupancy probabilities for selected energy levels within the bandgap.

Chapter 11

Carrier Concentration

In this chapter, we delve into the critical topic of determining carrier concentration in both intrinsic and extrinsic semiconductors. The concentration of charge carriers, namely electrons and holes, is a fundamental parameter that greatly influences the behavior and performance of semiconductor materials. To accurately assess and analyze the properties and characteristics of semiconductors, it is crucial to understand the methods for determining carrier concentration.

Intrinsic Carrier Concentration

Intrinsic semiconductors are those where the carrier concentration is solely determined by the thermal excitation of electrons from the valence band to the conduction band. The intrinsic carrier concentration, denoted as n_i, represents the equilibrium number of charge carriers in an undoped semiconductor at a specific temperature.

In the case of intrinsic semiconductors, analyzing carrier concentration involves considering the energy states available in the valence band and the conduction band. At thermal equilibrium, these energy states are occupied according to the Fermi-Dirac distribution function.

The intrinsic carrier concentration can be approximated using the principles of charge neutrality and the Fermi-Dirac distribution. For direct bandgap semiconductors, the intrinsic carrier concentration can be determined using the expression:

$$n_i = \left(\frac{2\pi m_e^* k_B T}{h^2}\right)^{3/2} \cdot e^{-\frac{E_g}{2k_B T}}$$

where m_e^* is the effective mass of an electron, h is the Planck constant, E_g is the energy bandgap, k_B is the Boltzmann constant, and T is the temperature. The intrinsic carrier concentration exponentially decreases with increasing energy bandgap and is positively correlated with temperature.

Extrinsic Carrier Concentration

Extrinsic semiconductors are those that have been intentionally doped with impurities to alter the carrier concentration. This process introduces additional energy states within the bandgap, leading to an increase in the concentration of either electrons (n-type) or holes (p-type).

The carrier concentration in extrinsic semiconductors is influenced by the doping level, the energy states introduced by the impurity atoms, and the thermal energy available at a given temperature. Mathematically, the carrier concentration in extrinsic semiconductors can be expressed as:

$$n = N_d + \frac{N_c}{N_v} \cdot \exp\left(\frac{E_c - E_F}{k_B T}\right)$$

for n-type doping, where N_d represents the donor concentration, N_c is the effective density of states in the conduction band, N_v is the effective density of states in the valence band, E_c is the energy of the conduction band edge, E_F is the Fermi energy level, k_B is the Boltzmann constant, and T denotes the temperature.

Similarly, for p-type doping, the carrier concentration is given by:

$$p = N_a + \frac{N_v}{N_c} \cdot \exp\left(\frac{E_F - E_v}{k_B T}\right)$$

where N_a represents the acceptor concentration, and E_v is the energy of the valence band edge.

1 Law of Mass Action

The Law of Mass Action is a vital principle governing the carrier concentration in doped semiconductors. According to this law, the product of the majority carrier concentration (either electrons or holes) and minority carrier concentration is constant at thermal equilibrium. Mathematically, this law can be expressed as:

$$n \cdot p = n_i^2$$

where n and p denote the majority and minority carrier concentrations, respectively, and n_i represents the intrinsic carrier concentration.

The Law of Mass Action provides insights into the relationship between majority and minority carrier concentrations and highlights the role of doping concentrations. It forms the basis for understanding the behavior of charge carriers in doped semiconductors and the functioning of various semiconductor devices.

Impurity Ionization and Compensation

In extrinsic semiconductors, impurity atoms introduced through doping can undergo ionization processes. At low temperatures, most dopant atoms remain ionized, resulting in a significant concentration of free carriers. However, with increased temperature, the ionization process becomes more favorable, leading to a decrease in the concentration of free carriers.

1 Compensation

Compensation is a phenomenon observed in doped semiconductors when impurity atoms of opposite doping types compensate for each other, resulting in a reduced net carrier concentration. This can occur when the concentration of one type of impurity exceeds the other, leading to the formation of donor-acceptor pairs.

The effect of compensation can be quantified using the compensation ratio, K, which is defined as the ratio of the acceptor concentration to the donor concentration. A compensation ratio greater than 1 indicates the presence of excess acceptor atoms, resulting in a decrease in the net carrier concentration.

Conclusion

The determination of carrier concentration in intrinsic and extrinsic semiconductors is crucial for understanding the behavior and electrical properties of semiconductor materials. Intrinsic carrier concentration is primarily influenced by the energy bandgap and temperature, while extrinsic carrier concentration is affected by the type and concentration of dopants. The Law of Mass Action provides further insights into the relationship between majority and minority carrier concentrations. Additionally, impurity ionization and compensation play essential roles in determining the net carrier concentration in extrinsic semiconductors.

The understanding and control of carrier concentration enable the design and optimization of semiconductor devices for various applications, ranging from electronic circuits to optoelectronics and beyond. By accurately determining and manipulating carrier concentration, researchers and engineers can unlock the full potential of semiconductors and drive advancements in technology.

Python Code Snippet

Below is a Python code snippet that calculates intrinsic and extrinsic carrier concentrations based on the equations and algorithms discussed in this chapter.

```python
import numpy as np

def intrinsic_carrier_concentration(Eg, m_e_star, T):
    '''
    Calculate the intrinsic carrier concentration for a
    ↪ semiconductor.
    :param Eg: Energy bandgap in eV.
    :param m_e_star: Effective mass of electron in kg.
    :param T: Temperature in Kelvin.
    :return: Intrinsic carrier concentration (n_i) in m^-3.
    '''
    k_B = 1.380649e-23  # Boltzmann constant in J/K
    h = 6.62607015e-34  # Planck's constant in J*s
    n_i = ((2 * np.pi * m_e_star * k_B * T) / h**2) ** (3/2) *
    ↪ np.exp(-Eg / (2 * k_B * T))
    return n_i

def extrinsic_n_type(N_d, E_c, E_F, T):
    '''
    Calculate carrier concentration for n-type semiconductor.
    :param N_d: Donor concentration in m^-3.
    :param E_c: Energy of the conduction band edge in eV.
    :param E_F: Fermi energy level in eV.
    :param T: Temperature in Kelvin.
    :return: Electron concentration (n) in m^-3.
```

```python
    '''
    k_B = 8.617333262145e-5  # Boltzmann constant in eV/K
    N_c = 1.0e25  # Effective density of states in the conduction
    ↪ band (example value)
    n = N_d + (N_c * np.exp((E_c - E_F) / (k_B * T)))
    return n

def extrinsic_p_type(N_a, E_F, E_v, T):
    '''
    Calculate carrier concentration for p-type semiconductor.
    :param N_a: Acceptor concentration in m^-3.
    :param E_F: Fermi energy level in eV.
    :param E_v: Energy of the valence band edge in eV.
    :param T: Temperature in Kelvin.
    :return: Hole concentration (p) in m^-3.
    '''
    k_B = 8.617333262145e-5  # Boltzmann constant in eV/K
    N_v = 1.0e25  # Effective density of states in the valence band
    ↪ (example value)
    p = N_a + (N_v * np.exp((E_F - E_v) / (k_B * T)))
    return p

def law_of_mass_action(n, p):
    '''
    Validate the Law of Mass Action for the semiconductor.
    :param n: Majority carrier concentration (electrons) in m^-3.
    :param p: Minority carrier concentration (holes) in m^-3.
    :return: Product of carrier concentrations (should equal n_i^2).
    '''
    n_i_squared = n * p
    return n_i_squared

# Constants for calculations
Eg = 1.12  # Energy bandgap for Silicon in eV
m_e_star = 9.11e-31  # Effective mass of an electron in kg
T = 300  # Temperature in Kelvin

# Intrinsic carrier concentration
n_i = intrinsic_carrier_concentration(Eg, m_e_star, T)

# Extrinsic carrier concentration example for n-type and p-type
N_d = 1e24  # Donor concentration in m^-3
E_c = 1.12  # Energy of conduction band edge in eV
E_F_n = 0.56  # Fermi energy level for n-type in eV

n_n_type = extrinsic_n_type(N_d, E_c, E_F_n, T)

N_a = 1e24  # Acceptor concentration in m^-3
E_v = 0.0  # Energy of valence band edge in eV
E_F_p = 0.4  # Fermi energy level for p-type in eV

p_p_type = extrinsic_p_type(N_a, E_F_p, E_v, T)
```

```
# Validate Law of Mass Action
mass_action_product = law_of_mass_action(n_n_type, p_p_type)

# Output results
print("Intrinsic Carrier Concentration (n_i):", n_i, "m^-3")
print("n-type Carrier Concentration (n):", n_n_type, "m^-3")
print("p-type Carrier Concentration (p):", p_p_type, "m^-3")
print("Mass Action Product (n * p):", mass_action_product, "m^-6")
print("n_i^2 (should match with n * p):", n_i**2, "m^-6")
```

This code defines four functions:

- **intrinsic_carrier_concentration** calculates the intrinsic carrier concentration based on energy bandgap, effective mass, and temperature.
- **extrinsic_n_type** computes the electron concentration for n-type semiconductors given donor concentration, conduction band edge, and Fermi level.
- **extrinsic_p_type** calculates the hole concentration for p-type semiconductors based on acceptor concentration, valence band edge, and Fermi level.
- **law_of_mass_action** validates the Law of Mass Action for given majority and minority carrier concentrations.

The provided example calculates the intrinsic, n-type, and p-type carrier concentrations for a silicon semiconductor and validates the Law of Mass Action by comparing the product of the carrier concentrations with n_i^2. The results are then printed to the console.

Chapter 12

Poisson's Equation

In this chapter, we will explore the role of Poisson's equation in semiconductor device physics, with a particular focus on its significance in pn junctions. Poisson's equation is a fundamental equation in electrostatics that plays a crucial role in understanding the behavior and characteristics of semiconductor devices.

Introduction to Poisson's Equation

Poisson's equation is a partial differential equation that relates the distribution of electric potential, $\Phi(\mathbf{r})$, to the charge density, $\rho(\mathbf{r})$, within a given region of space. Mathematically, it can be expressed as:

$$\nabla^2 \Phi(\mathbf{r}) = -\frac{\rho(\mathbf{r})}{\varepsilon}$$

where ∇^2 is the Laplacian operator, $\mathbf{r}$ represents the position vector, and ε denotes the permittivity of the material.

Role in Semiconductor Device Physics

In the context of semiconductor device physics, Poisson's equation is of utmost importance in analyzing and designing various electronic devices, including pn junctions. The equation allows us to determine the electric potential distribution and, consequently, the electric field within the device.

1 Poisson's Equation in pn Junctions

A pn junction is formed by joining two regions of different doping types in a semiconductor material, namely the p-region (with excess holes) and the n-region (with excess electrons). Poisson's equation is a key component in understanding the behavior of pn junctions.

Depletion Region

One significant aspect of a pn junction is the depletion region, which is formed at the interface between the p- and n-regions due to the diffusion of charge carriers. In this region, Poisson's equation allows us to derive the depletion region width and the electric field distribution.

By solving Poisson's equation under appropriate boundary conditions, we can determine the potential distribution across the pn junction and the resulting electric field. This field plays a critical role in governing the movement of charge carriers, particularly the drift of minority carriers across the junction.

Built-in Potential

Another important feature of a pn junction is the built-in potential, which refers to the electric potential difference across the junction in thermal equilibrium. It is directly influenced by the doping concentrations on either side of the junction. Poisson's equation enables the calculation of the built-in potential by considering the charge distribution and permittivity within the pn junction.

The built-in potential establishes a barrier for the movement of charge carriers and determines the behavior of the junction under different biasing conditions.

2 Boundary Conditions

Solving Poisson's equation in pn junctions requires appropriate boundary conditions that describe the physical characteristics of the junction. These conditions relate to the continuity of electric potential and the electric field at the junction interface.

At the pn junction interface, the continuity of electric potential specifies that the potential on both sides of the interface must be equal. Mathematically, this can be expressed as:

$$\Phi_p(\mathbf{r}) = \Phi_n(\mathbf{r})$$

where $\Phi_p(\mathbf{r})$ represents the potential in the p-region and $\Phi_n(\mathbf{r})$ denotes the potential in the n-region.

Similarly, the continuity of the normal component of the electric field at the interface requires that the electric field is continuous across the junction. This can be expressed as:

$$-\varepsilon_p \frac{d\Phi_p}{dx} = -\varepsilon_n \frac{d\Phi_n}{dx}$$

where ε_p and ε_n denote the permittivity of the p- and n-regions, respectively, and $\frac{d\Phi}{dx}$ represents the derivative of electric potential with respect to the distance across the junction.

Conclusion

Poisson's equation serves as a fundamental tool in semiconductor device physics, especially in the analysis of pn junctions. By solving Poisson's equation, we can determine the potential distribution, electric field, and other electrical characteristics within the junction. The equation enables us to study the depletion region and the built-in potential, both of which are critical parameters in understanding the behavior of pn junctions.

The appropriate application of Poisson's equation, along with suitable boundary conditions, allows researchers and engineers to gain valuable insights into the operation and optimization of various semiconductor devices. Through a deep understanding of Poisson's equation, we can unlock the potential of pn junctions and devise innovative solutions for a wide range of electronic applications.

Python Code Snippet

Below is a Python code snippet that implements important equations and algorithms related to Poisson's equation and pn junction analysis.

```python
import numpy as np
import matplotlib.pyplot as plt

def poisson_equation_solve(charge_density, epsilon, length,
    num_points):
```

```python
    '''
    Solve Poisson's equation using finite difference method.
    :param charge_density: Charge density as a numpy array.
    :param epsilon: Permittivity of the material.
    :param length: Length of the region in meters.
    :param num_points: Number of discrete points in the region.
    :return: Electric potential distribution as a numpy array.
    '''
    dx = length / (num_points - 1)
    potential = np.zeros(num_points)

    # Construct the right-hand side based on charge density
    rhs = -charge_density / epsilon

    # Finite Difference Method
    for i in range(1, num_points - 1):
        potential[i] = (potential[i - 1] + potential[i + 1] + dx**2
        ↪   * rhs[i]) / 2

    return potential

def depletion_region_width(doping_p, doping_n, epsilon):
    '''
    Calculate the width of the depletion region in a pn junction.
    :param doping_p: Doping concentration of p-region (cm^-3).
    :param doping_n: Doping concentration of n-region (cm^-3).
    :param epsilon: Permittivity of the semiconductor material
    ↪   (F/m).
    :return: Width of the depletion region in meters.
    '''
    N_a = doping_p  # Acceptor concentration
    N_d = doping_n  # Donor concentration
    V_bi = (0.0259 * np.log((N_a * N_d) ** 0.5))  # Built-in
    ↪   potential in volts
    w = np.sqrt((2 * epsilon * V_bi) / (q * (N_a + N_d)))  #
    ↪   Depletion width equation
    return w

def built_in_potential(doping_p, doping_n):
    '''
    Calculate the built-in potential of a pn junction.
    :param doping_p: Doping concentration of p-region (cm^-3).
    :param doping_n: Doping concentration of n-region (cm^-3).
    :return: Built-in potential in volts.
    '''
    kT = 0.0259  # Thermal voltage at room temperature in volts
    V_bi = kT * np.log((doping_p * doping_n) / (ni**2))  # ni =
    ↪   intrinsic carrier concentration
    return V_bi

# Constants and parameters
epsilon = 11.7 * 8.854187817e-12  # Permittivity of silicon (F/m)
q = 1.6e-19  # Charge of an electron (C)
```

```python
ni = 1.5e10    # Intrinsic carrier concentration for silicon (cm^-3)

doping_p = 1e16    # Doping concentration of p-region (cm^-3)
doping_n = 1e16    # Doping concentration of n-region (cm^-3)

length = 1e-6    # Length of region in meters
num_points = 100    # Number of points for potential distribution

# Charge density example: uniform for simplicity
charge_density = np.zeros(num_points)
charge_density[int(num_points/4):int(num_points/2)] = doping_p    #
   P-type region
charge_density[int(num_points/2):int(3*num_points/4)] = -doping_n    #
   N-type region

# Solve Poisson's equation
potential = poisson_equation_solve(charge_density, epsilon, length,
   num_points)

# Calculate depletion width and built-in potential
depletion_width = depletion_region_width(doping_p, doping_n,
   epsilon)
built_in_pot = built_in_potential(doping_p, doping_n)

# Plotting the potential distribution
x = np.linspace(0, length, num_points)
plt.plot(x, potential, label='Electric Potential')
plt.xlabel('Position (m)')
plt.ylabel('Potential (V)')
plt.title('Potential Distribution in a PN Junction')
plt.axvline(x=depletion_width, color='r', linestyle='--',
   label='Depletion Width')
plt.legend()
plt.grid()
plt.show()

# Output results
print("Depletion Width:", depletion_width, "m")
print("Built-in Potential:", built_in_pot, "V")
```

This code defines three functions:

- `poisson_equation_solve` solves Poisson's equation using the finite difference method to obtain the electric potential distribution from the charge density.
- `depletion_region_width` calculates the width of the depletion region in a pn junction based on doping concentrations and permittivity.
- `built_in_potential` computes the built-in potential of a pn junction.

The provided example calculates the potential distribution within a pn junction, plots the electric potential, and prints the depletion width and built-in potential, showcasing the key principles discussed in the chapter.

Chapter 13

Continuity Equation

The continuity equation plays a crucial role in the analysis of charge carriers in semiconductors under non-static conditions. It provides a mathematical statement of the conservation of charge, allowing us to understand the dynamics and behavior of carriers in semiconductor devices. In this chapter, we will derive the continuity equation and explore its implications in semiconductor physics.

Derivation of the Continuity Equation

Consider a localized region within a semiconductor material, where charge carriers are moving under the influence of an electric field. The density of mobile charge carriers in this region can be denoted as $n(\mathbf{r},t)$ for electrons and $p(\mathbf{r},t)$ for holes, where $\mathbf{r}$ represents the position vector and t denotes time. The continuity equation for each carrier type can be derived based on the principle of conservation of charge.

1 Conservation of Charge

To derive the continuity equation, we start with the principle of conservation of charge, which states that the rate of change of charge within a given region must equal the net charge flow into or out of the region.

For electrons, the rate of change of charge density is given by

$$\frac{\partial n}{\partial t}$$

Considering the flux of electrons due to their motion, the net charge flow can be expressed as

$$-\nabla \cdot \mathbf{J}_n$$

where $\mathbf{J}_n$ is the electron current density. By equating the rate of change of charge density to the net charge flow, we obtain the continuity equation for electrons:

$$\frac{\partial n}{\partial t} = -\nabla \cdot \mathbf{J}_n$$

Similarly, the continuity equation for holes can be derived as:

$$\frac{\partial p}{\partial t} = -\nabla \cdot \mathbf{J}_p$$

where $\mathbf{J}_p$ is the hole current density. These coupled partial differential equations describe the dynamics of charge carriers in semiconductors.

2 Current Density and Carrier Motion

To further understand the continuity equation, we express the current densities $\mathbf{J}_n$ and $\mathbf{J}_p$ in terms of carrier motion. For electrons, the current density can be written as:

$$\mathbf{J}_n = q\mu_n n \mathbf{E} + qD_n \nabla n$$

where q represents the charge of an electron, μ_n denotes the electron mobility, $\mathbf{E}$ represents the electric field, and D_n is the electron diffusion coefficient. The first term corresponds to the drift of electrons under the influence of the electric field, while the second term accounts for the diffusion of electrons due to concentration gradients.

Similarly, for holes, the current density is given by:

$$\mathbf{J}_p = -q\mu_p p \mathbf{E} + qD_p \nabla p$$

where μ_p represents the hole mobility and D_p denotes the hole diffusion coefficient. The first term corresponds to the drift of holes, while the second term accounts for the diffusion of holes.

3 Simplification and Final Form

By substituting the expressions for the current densities into the continuity equations, we can simplify and rewrite them in a more compact form. The continuity equation for electrons becomes:

$$\frac{\partial n}{\partial t} = -\nabla \cdot (q\mu_n n \mathbf{E}) - \nabla \cdot (qD_n \nabla n)$$

Similarly, the continuity equation for holes can be written as:

$$\frac{\partial p}{\partial t} = -\nabla \cdot (-q\mu_p p \mathbf{E}) - \nabla \cdot (qD_p \nabla p)$$

These equations represent the conservation of charge for electrons and holes in semiconductors under non-static conditions.

Implications and Applications

The continuity equation has significant implications in semiconductor physics and is employed in various device models and simulations. By solving the continuity equations, we can analyze the behavior of charge carriers in semiconductors under different operating conditions, such as biased pn junctions or transistor operation.

The continuity equation allows us to understand the flow and redistribution of charge carriers in response to external stimuli, such as applied voltages or temperature changes. It serves as a fundamental tool in developing accurate models for semiconductor devices, aiding in their design and optimization.

Moreover, the continuity equation forms the basis for many other important equations and concepts in semiconductor physics, such as the drift-diffusion model and the analysis of carrier transport.

Conclusion

The continuity equation is a fundamental equation in semiconductor physics, providing a mathematical description of the conservation of charge for electrons and holes. By solving the continuity equations, we can analyze the dynamics of charge carriers in semiconductors under non-static conditions and gain insights into their behavior.

With its implications and applications in device modeling and simulations, the continuity equation serves as a cornerstone in the study and understanding of semiconductor devices. It paves the way for further exploration of carrier transport mechanisms and the development of advanced device concepts."'latex

Python Code Snippet

Below is a Python code snippet that calculates the current density for electrons and holes, as well as the rate of change of carrier concentration according to the continuity equations discussed in this chapter.

```python
def calculate_current_density(n, p, E, mu_n, mu_p, D_n, D_p, grad_n,
    grad_p, q):
    '''
    Calculate the current density for electrons and holes.
    :param n: Electron concentration (m^-3).
    :param p: Hole concentration (m^-3).
    :param E: Electric field (V/m).
    :param mu_n: Electron mobility (m^2/V*s).
    :param mu_p: Hole mobility (m^2/V*s).
    :param D_n: Electron diffusion coefficient (m^2/s).
    :param D_p: Hole diffusion coefficient (m^2/s).
    :param grad_n: Gradient of electron concentration.
    :param grad_p: Gradient of hole concentration.
    :param q: Charge of an electron (C).
    :return: Current density for electrons and holes (A/m^2).
    '''
    J_n = q * (mu_n * n * E + D_n * grad_n)
    J_p = -q * (mu_p * p * E + D_p * grad_p)
    return J_n, J_p

def continuity_equation_rate_of_change(n, p, J_n, J_p, volume):
    '''
    Calculate the rate of change of carrier concentration.
    :param n: Electron concentration (m^-3).
    :param p: Hole concentration (m^-3).
    :param J_n: Electron current density (A/m^2).
    :param J_p: Hole current density (A/m^2).
    :param volume: Volume of the region (m^3).
    :return: Rate of change of electron and hole concentrations
        (m^-3/s).
    '''
    rate_change_n = -J_n / volume
    rate_change_p = -J_p / volume

    return rate_change_n, rate_change_p
```

```python
# Input parameters
q = 1.6e-19    # Charge of an electron in coulombs
n = 1e20       # Electron concentration in m^-3
p = 1e20       # Hole concentration in m^-3
E = 1000       # Electric field in V/m
mu_n = 0.15    # Electron mobility in m^2/V*s
mu_p = 0.05    # Hole mobility in m^2/V*s
D_n = 0.025    # Electron diffusion coefficient in m^2/s
D_p = 0.01     # Hole diffusion coefficient in m^2/s
grad_n = 1e-6  # Gradient of electron concentration
grad_p = 1e-6  # Gradient of hole concentration
volume = 1e-6  # Volume of the region in m^3

# Calculations
J_n, J_p = calculate_current_density(n, p, E, mu_n, mu_p, D_n, D_p,
↪   grad_n, grad_p, q)
rate_change_n, rate_change_p = continuity_equation_rate_of_change(n,
↪   p, J_n, J_p, volume)

# Output results
print("Electron Current Density:", J_n, "A/m^2")
print("Hole Current Density:", J_p, "A/m^2")
print("Rate of Change of Electron Concentration:", rate_change_n,
↪   "m^-3/s")
print("Rate of Change of Hole Concentration:", rate_change_p,
↪   "m^-3/s")
```

This code defines two functions:

- `calculate_current_density` calculates the current density for electrons and holes based on their concentrations, electric field, mobilities, diffusion coefficients, and concentration gradients.
- `continuity_equation_rate_of_change` computes the rate of change of carrier concentration for electrons and holes using the calculated current densities.

The example provided calculates the current densities and rates of change of carrier concentrations and then prints the results, aiding in the analysis of semiconductor dynamics. "'

Chapter 14

Drift-Diffusion Model

The drift-diffusion model is a fundamental approach to modeling charge transport in semiconductors, providing insights into the dynamics and behavior of charge carriers. This chapter explores the charge transport mechanisms via drift and diffusion, shedding light on the mathematical formulation and implications of this model.

Drift and Diffusion

In semiconductors, charge carriers, such as electrons and holes, can move in response to internal or external electric fields. The charge transport occurs through two main mechanisms: drift and diffusion.

1 Drift

Drift refers to the movement of charge carriers in response to an applied electric field. When an electric field $\mathbf{E}$ is present, the charge carriers experience a force given by $q\mathbf{E}$, where q is the charge of the carrier. This force causes the carriers to accelerate, leading to a net motion or drift in the direction of the electric field. The carrier mobility μ describes the ease with which carriers can move in response to the applied electric field.

For electrons, the drift current density $\mathbf{J}_{n,\text{drift}}$ is given by:

$$\mathbf{J}_{n,\text{drift}} = q\mu_n n \mathbf{E}$$

where n represents the electron concentration and μ_n is the electron mobility. Similarly, for holes, the drift current density $\mathbf{J}_{p,\text{drift}}$ is given by:

$$\mathbf{J}_{p,\text{drift}} = -q\mu_p p \mathbf{E}$$

where p denotes the hole concentration and μ_p represents the hole mobility. The drift current arises from the net drift of charge carriers in response to the electric field.

2 Diffusion

Diffusion, on the other hand, arises due to the concentration gradient of charge carriers. In regions where there is a variation in carrier concentration, charge carriers tend to move from regions of higher concentration to regions of lower concentration. This movement is driven by the tendency to equalize the carrier concentrations, similar to the diffusion of a gas from an area of high concentration to an area of low concentration. Diffusion is described by the diffusion coefficient D.

The diffusion current density $\mathbf{J}_{\text{diff}}$ for both electrons and holes can be expressed as:

$$\mathbf{J}_{\text{diff}} = qD\nabla n$$

where n represents the carrier concentration and ∇n denotes the gradient of the concentration. The diffusion current arises from the concentration gradient, driving the carriers from regions of higher concentration to regions of lower concentration.

Drift-Diffusion Equations

By considering both drift and diffusion mechanisms, we arrive at the drift-diffusion model, which consists of a set of coupled partial differential equations describing the dynamics of the charge carriers. The drift-diffusion equations capture the balance between the drift and diffusion currents, as well as the rate of change of carrier concentration.

For electrons, the drift-diffusion equation can be written as:

$$\frac{\partial n}{\partial t} = -\nabla \cdot \mathbf{J}_{n,\text{drift}} - \nabla \cdot \mathbf{J}_{\text{diff}} + G_n - R_n$$

where $\frac{\partial n}{\partial t}$ represents the rate of change of the electron concentration n with respect to time t, $\nabla \cdot \mathbf{J}_{n,\text{drift}}$ represents the divergence of the drift current density, $\nabla \cdot \mathbf{J}_{\text{diff}}$ represents the divergence of the diffusion current density, G_n represents the generation of electron-hole pairs, and R_n represents the recombination of electron-hole pairs.

Similarly, the drift-diffusion equation for holes can be expressed as:

$$\frac{\partial p}{\partial t} = -\nabla \cdot \mathbf{J}_{p,\text{drift}} - \nabla \cdot \mathbf{J}_{\text{diff}} + G_p - R_p$$

where $\frac{\partial p}{\partial t}$ represents the rate of change of the hole concentration p with respect to time t, $\nabla \cdot \mathbf{J}_{p,\text{drift}}$ represents the divergence of the drift current density for holes, $\nabla \cdot \mathbf{J}_{\text{diff}}$ represents the divergence of the diffusion current density, G_p represents the generation of electron-hole pairs, and R_p represents the recombination of electron-hole pairs.

Implications and Applications

The drift-diffusion model is a powerful tool for investigating the behavior of charge carriers in various semiconductor devices, such as pn junctions and transistors. By solving the drift-diffusion equations, we can gain insights into the current-voltage characteristics, transient response, and steady-state behavior of these devices.

Moreover, the drift-diffusion model enables the analysis of carrier transport and the understanding of device performance under different operating conditions, such as temperature variations and doping profiles. This model allows for the optimization and design of semiconductor devices based on the desired performance characteristics.

The significance of the drift-diffusion model extends beyond the device level. It forms the foundation for more advanced models, such as the hydrodynamic model and the ensemble Monte Carlo method, which capture additional physics, such as high-field effects and quantum effects.

In summary, the drift-diffusion model provides a robust framework for studying charge transport in semiconductors, offering valuable insights into the behavior and performance of semiconductor devices. Its applications span a wide range of fields, from device engineering to materials science, guiding advancements in semiconductor technology.

Python Code Snippet

Below is a Python code snippet that implements the important equations and algorithms discussed in the Drift-Diffusion Model chapter. The code calculates the drift and diffusion current densities, applies the drift-diffusion equations, and simulates carrier concentration change over time.

```python
import numpy as np
import matplotlib.pyplot as plt

def calculate_drift_current(electric_field, electron_concentration,
    electron_mobility):
    '''
    Calculate the drift current density for electrons.
    :param electric_field: Applied electric field in volts/meter.
    :param electron_concentration: Concentration of electrons in
        m^-3.
    :param electron_mobility: Electron mobility in m^2/(V.s).
    :return: Drift current density for electrons in A/m^2.
    '''
    q = 1.602e-19  # Charge of an electron in coulombs
    return q * electron_mobility * electron_concentration *
        electric_field


def calculate_diffusion_current(diffusion_coefficient,
    carrier_concentration_gradient):
    '''
    Calculate the diffusion current density.
    :param diffusion_coefficient: Diffusion coefficient in m^2/s.
    :param carrier_concentration_gradient: Gradient of carrier
        concentration in m^-4.
    :return: Diffusion current density in A/m^2.
    '''
    q = 1.602e-19  # Charge of an electron in coulombs
    return q * diffusion_coefficient *
        carrier_concentration_gradient


def drift_diffusion_simulation(time_steps, spatial_steps,
    electric_field, initial_concentration, electron_mobility,
    diffusion_coefficient):
    '''
    Run a simple simulation of the drift-diffusion model.
    :param time_steps: Number of time steps for the simulation.
```

```
    :param spatial_steps: Number of spatial steps for the
    ↪  simulation.
    :param electric_field: Applied electric field in volts/meter.
    :param initial_concentration: Initial concentration of charge
    ↪  carriers in m^-3.
    :param electron_mobility: Electron mobility in m^2/(V.s).
    :param diffusion_coefficient: Diffusion coefficient in m^2/s.
    :return: Array of carrier concentrations over time.
    '''
    concentrations = np.zeros((time_steps, spatial_steps))
    concentrations[0, :] = initial_concentration

    for t in range(1, time_steps):
        for x in range(1, spatial_steps-1):
            # Calculate the concentration gradient
            gradient = concentrations[t-1, x+1] -
            ↪  concentrations[t-1, x-1]
            diffusion_current =
            ↪  calculate_diffusion_current(diffusion_coefficient,
            ↪  gradient)
            drift_current = calculate_drift_current(electric_field,
            ↪  concentrations[t-1, x], electron_mobility)

            # Update carrier concentration based on drift and
            ↪  diffusion
            concentrations[t, x] = concentrations[t-1, x] +
            ↪  (diffusion_current + drift_current)

    return concentrations


# Parameters for the simulation
time_steps = 100      # total time steps
spatial_steps = 50    # spatial grid points
initial_concentration = 1e20   # m^-3
electric_field = 1000          # Volts/meter
electron_mobility = 0.14       # m^2/(V.s) for silicon
diffusion_coefficient = 36e-4  # m^2/s for silicon

# Run the simulation
carrier_concentration = drift_diffusion_simulation(time_steps,
↪  spatial_steps, electric_field, initial_concentration,
↪  electron_mobility, diffusion_coefficient)

# Visualize the results
plt.figure(figsize=(10, 5))
for t in range(0, time_steps, 10):
    plt.plot(carrier_concentration[t, :], label=f'Time step {t}')
plt.title('Carrier Concentration Over Time')
plt.xlabel('Spatial Index')
plt.ylabel('Carrier Concentration (m^-3)')
plt.legend()
plt.grid()
```

```
plt.show()
```

This code defines three functions:

- `calculate_drift_current` computes the drift current density of electrons based on the electric field, carrier concentration, and mobility.
- `calculate_diffusion_current` calculates the diffusion current density based on the concentration gradient and diffusion coefficient.
- `drift_diffusion_simulation` implements a simple simulation of charge carrier transport by solving the drift-diffusion equations over specified time and spatial grids.

The simulation visualizes the carrier concentration changes over time in a spatial domain under a constant electric field, allowing insights into the transport dynamics of carriers in semiconductors.

Chapter 15

Einstein Relation

In this chapter, we explore an essential relationship in semiconductor physics known as the Einstein relation. The Einstein relation establishes a connection between the diffusion coefficient and the carrier mobility, providing valuable insights into the behavior of charge carriers in semiconductors. We will delve into the derivation and implications of this relation, shedding light on its significance in understanding semiconductor transport phenomena.

Derivation of the Einstein Relation

The Einstein relation arises from a fundamental concept in statistical physics known as the equilibrium condition. In thermal equilibrium, the average drift velocity of a charged particle is zero due to the balance between the electric force and the particle's thermal energy. By considering this equilibrium condition, we can derive the Einstein relation.

Let us consider a charged particle in a semiconductor under the influence of both an external electric field and thermal fluctuations. The motion of the particle can be described by the Langevin equation:

$$\frac{dv}{dt} = \frac{qE}{m} - \frac{v}{\tau} + \sqrt{\frac{2k_B T}{m}} \eta(t)$$

where v represents the velocity of the particle, q is its charge, E denotes the applied electric field, m represents the particle mass, τ is the relaxation time characterizing the momentum scattering,

k_B is Boltzmann's constant, T represents the temperature, and $\eta(t)$ denotes a white Gaussian noise term representing the thermal fluctuations.

Analyzing the equilibrium condition where the average drift velocity is zero, we can neglect the time derivative in the Langevin equation. This yields:

$$0 = \frac{qE}{m} - \frac{v}{\tau}$$

Simplifying, we find that the average drift velocity $\bar{v}$ is given by:

$$\bar{v} = \frac{qE\tau}{m}$$

In semiconductor physics, the average drift velocity is related to the carrier mobility μ through the expression:

$$\bar{v} = \mu E$$

Combining the above equations, we obtain the Einstein relation:

$$\mu = \frac{q\tau}{m}$$

This relation provides a direct connection between the carrier mobility, transport relaxation time, and the charge-to-mass ratio of the carriers in a semiconductor.

Implications of the Einstein Relation

The Einstein relation has profound implications for the understanding of charge transport in semiconductors. By relating the carrier mobility to the transport relaxation time, we gain insights into the factors influencing carrier motion and diffusion in semiconductor materials.

The Einstein relation emphasizes that the mobility of carriers is directly proportional to their diffusion coefficient. This connection demonstrates that the fluctuations induced by thermal energy play a crucial role in determining the overall mobility and diffusion behavior of carriers.

Additionally, the Einstein relation reveals the dependency of mobility on carrier scattering phenomena, which affect the transport relaxation time. Scattering processes, such as phonon scattering, impurity scattering, and carrier-carrier interactions, influence the carrier mobility and diffusion in semiconductors.

Furthermore, the Einstein relation allows for the estimation of diffusion coefficients based on carrier mobility measurements, and vice versa. This relationship provides a convenient means of obtaining or verifying important transport properties of semiconductor materials.

In summary, the Einstein relation provides a valuable link between the diffusion coefficient and carrier mobility, shedding light on the intricate interplay between thermal fluctuations, scattering mechanisms, and charge transport in semiconductors. Understanding and applying this relation are crucial for the comprehensive analysis and design of semiconductor devices.

"'latex

Python Code Snippet

Below is a Python code snippet that calculates the carrier mobility and diffusion coefficient in semiconductors, along with their relation through the Einstein relation.

```python
def calculate_mobility(charge, mass, relaxation_time):
    '''
    Calculate the carrier mobility in semiconductors.
    :param charge: Charge of the carrier in coulombs.
    :param mass: Mass of the carrier in kilograms.
    :param relaxation_time: Average time between collisions in
     ↪ seconds.
    :return: Mobility in m^2/(V·s).
    '''
    mobility = (charge * relaxation_time) / mass
    return mobility


def calculate_diffusion_coefficient(mobility, temperature):
    '''
    Calculate the diffusion coefficient based on mobility and
     ↪ temperature using Einstein relation.
    :param mobility: Carrier mobility in m^2/(V·s).
    :param temperature: Temperature in Kelvin.
    :return: Diffusion coefficient in m^2/s.
    '''
    k_B = 1.380649e-23  # Boltzmann constant in J/K
    diffusion_coefficient = mobility * (k_B * temperature)
    return diffusion_coefficient
```

```python
# Inputs for calculations
charge = 1.602176634e-19  # Charge of an electron in coulombs
mass = 9.10938356e-31  # Mass of an electron in kilograms
relaxation_time = 1e-14  # Relaxation time in seconds
temperature = 300  # Temperature in Kelvin (room temperature)

# Calculations
mobility = calculate_mobility(charge, mass, relaxation_time)
diffusion_coefficient = calculate_diffusion_coefficient(mobility,
    temperature)

# Output results
print("Carrier Mobility:", mobility, "m^2/(V·s)")
print("Diffusion Coefficient:", diffusion_coefficient, "m^2/s")
```

This code defines two functions:

- `calculate_mobility` calculates the mobility of charge carriers in a semiconductor given the charge, mass, and relaxation time.
- `calculate_diffusion_coefficient` computes the diffusion coefficient using the carrier mobility and temperature, applying the Einstein relation.

The example provided calculates the carrier mobility and diffusion coefficient for an electron at room temperature and prints the results. "'

Chapter 16

PN Junctions

PN junctions are one of the most fundamental and widely used components in semiconductor devices. They form a crucial building block for various electronic devices, including diodes, transistors, and integrated circuits. In this chapter, we delve into the analysis of PN junctions, with a specific focus on the depletion approximation and the concept of built-in potential. These concepts are essential for understanding the behavior of PN junctions and their applications in electronic devices.

Depletion Approximation

The depletion region in a PN junction is a crucial region where the concentration of majority carriers is significantly reduced due to the diffusion of minority carriers across the junction. The analysis of the PN junction under equilibrium conditions involves making simplifying assumptions to simplify the mathematical model. The depletion approximation is a widely-used technique that simplifies the analysis of the PN junction by assuming that the concentration of majority carriers in the depletion region is zero.

To mathematically describe the depletion approximation, we consider a one-dimensional PN junction in thermal equilibrium. Let x denote the spatial coordinate across the junction, and N_A and N_D represent the acceptor and donor doping concentrations, respectively. The built-in potential V_0 across the junction creates an electric field that results in the redistribution of charge carriers and formation of the depletion region.

Within the depletion region, the concentration of electrons ($n(x)$) and holes ($p(x)$) can be approximated as zero due to recombination processes and the diffusion of majority carriers. Outside the depletion region, the concentrations of electrons and holes can be described by their equilibrium concentrations in the uniform P and N regions.

The depletion width (W_{depl}), which represents the region depleted of majority carriers, can be calculated using the relation:

$$W_{\text{depl}} = \sqrt{\frac{2\epsilon_s}{q}\left(\frac{1}{N_A} + \frac{1}{N_D}\right)V_0}$$

where ϵ_s is the permittivity of the semiconductor, and q denotes the elementary charge.

Built-in Potential

The built-in potential (V_0) is a crucial parameter that determines the behavior and characteristics of a PN junction. It arises due to the diffusion of minority carriers across the junction, leading to the formation of the depletion region. The built-in potential creates an electric field that acts as a barrier to the further diffusion of majority carriers, thus establishing a stable equilibrium in the PN junction.

The built-in potential can be mathematically derived by considering the equilibrium condition. Under equilibrium, the Fermi levels of the P and N regions align, resulting in a potential difference equal to the built-in potential across the junction. This potential difference can be expressed as:

$$V_0 = \frac{k_B T}{q} \ln\left(\frac{N_A N_D}{n_i^2}\right)$$

where k_B represents Boltzmann's constant, T denotes the temperature, and n_i represents the intrinsic carrier concentration.

The built-in potential plays a critical role in determining various characteristics of the PN junction, including the junction capacitance, the voltage-current relationship, and the behavior under external bias. Moreover, the built-in potential affects the formation of other important junctions, such as metal-oxide-semiconductor (MOS) capacitors and heterojunctions.

In summary, the depletion approximation and built-in potential are essential concepts in the analysis of PN junctions. The depletion approximation simplifies the mathematical model by assuming zero concentration of majority carriers within the depletion region. The built-in potential arises due to the diffusion of minority carriers and establishes a stable equilibrium in the PN junction. Understanding these concepts is crucial for comprehending the behavior and characteristics of PN junctions, enabling the design and analysis of a wide range of semiconductor devices.Sure! Below is a comprehensive Python code snippet demonstrating the important equations and formulas discussed in the chapter about PN junctions. The code includes functions for calculating the depletion width and built-in potential of a PN junction. It is presented in LaTeX using the minted package.

"'latex

Python Code Snippet

Below is a Python code snippet that calculates the depletion width and built-in potential for a PN junction based on the equations discussed in this chapter.

```
def calculate_depletion_width(N_A, N_D, V_0, epsilon_s, q):
    '''
    Calculate the depletion width of a PN junction.
    :param N_A: Acceptor doping concentration (m^-3).
    :param N_D: Donor doping concentration (m^-3).
    :param V_0: Built-in potential (V).
    :param epsilon_s: Permittivity of the semiconductor (F/m).
    :param q: Elementary charge (C).
    :return: Depletion width (m).
    '''
    return (2 * epsilon_s / q * (1/N_A + 1/N_D) * V_0) ** 0.5


def calculate_built_in_potential(N_A, N_D, n_i, k_B, T, q):
    '''
    Calculate the built-in potential of a PN junction.
    :param N_A: Acceptor doping concentration (m^-3).
    :param N_D: Donor doping concentration (m^-3).
    :param n_i: Intrinsic carrier concentration (m^-3).
    :param k_B: Boltzmann's constant (J/K).
    :param T: Temperature (K).
    :param q: Elementary charge (C).
    :return: Built-in potential (V).
    '''
```

```python
    return (k_B * T / q) * math.log((N_A * N_D) / (n_i ** 2))

# Constants and example values
import math

# Doping concentrations (in m^-3)
N_A = 1e25  # Example Acceptor concentration (P-type)
N_D = 1e24  # Example Donor concentration (N-type)

# Material properties
epsilon_s = 11.7 * 8.854187e-12  # Example for silicon in F/m
q = 1.602176634e-19  # Elementary charge in coulombs

# Temperature (in Kelvin)
T = 300  # Room temperature

# Intrinsic carrier concentration for silicon (m^-3)
n_i = 1.5e16  # Example for silicon

# Boltzmann's constant (J/K)
k_B = 1.380649e-23  # in J/K

# Calculate built-in potential
V_0 = calculate_built_in_potential(N_A, N_D, n_i, k_B, T, q)

# Calculate depletion width
W_depl = calculate_depletion_width(N_A, N_D, V_0, epsilon_s, q)

# Output results
print("Built-in Potential (V):", V_0)
print("Depletion Width (m):", W_depl)
```

This code defines two functions:

- `calculate_depletion_width` computes the width of the depletion region in a PN junction based on doping concentrations, built-in potential, permittivity, and elementary charge.
- `calculate_built_in_potential` calculates the built-in potential of a PN junction using the acceptor and donor concentrations, intrinsic carrier concentration, temperature, and physical constants.

The provided example calculates the built-in potential and depletion width for a PN junction and then prints the results. "'

Make sure to include the appropriate LaTeX packages in your document preamble to use the 'minted' environment correctly.

Chapter 17

Shockley Equation

The Shockley equation is a fundamental equation in semiconductor physics that describes the current-voltage characteristics of a pn junction. It provides insight into the behavior of the junction under different biasing conditions, allowing for the analysis and design of various semiconductor devices. In this chapter, we delve into the derivation and implications of the Shockley equation, providing a rigorous mathematical treatment and discussing its practical applications.

Derivation of the Shockley Equation

The Shockley equation is derived by applying the principles of carrier transport and recombination to a pn junction under forward bias. Consider a one-dimensional pn junction with a voltage bias applied across it. The voltage bias creates a potential difference ΔV across the junction, causing the majority carriers to drift and the minority carriers to diffuse. Under forward bias, the positive terminal of the voltage source is connected to the p-side, and the negative terminal is connected to the n-side.

To derive the Shockley equation, we start by considering the current density (J) flowing across the junction. The total current density can be expressed as the sum of the drift current density (J_{drift}) and the diffusion current density (J_{diff}). The drift current is due to the movement of majority carriers under the influence of the electric field, while the diffusion current arises from the concentration gradient of minority carriers.

The drift current density can be described using Fick's law, which relates the current density to the mobility (μ) and carrier concentration (n). For electrons in the n-side, the drift current density (J_{driftn}) can be written as:

$$J_{\text{driftn}} = q n \mu_n \frac{\Delta V}{L_n}$$

where q is the elementary charge and L_n is the length of the n-side.

Similarly, for holes in the p-side, the drift current density (J_{driftp}) can be expressed as:

$$J_{\text{driftp}} = -q p \mu_p \frac{\Delta V}{L_p}$$

where p denotes the hole concentration and μ_p represents the hole mobility. The negative sign arises from the opposite direction of hole movement.

The diffusion current density is given by the Einstein relation, which relates the current density to the diffusion coefficient (D) and concentration gradient. For electrons, the diffusion current density (J_{diffn}) can be written as:

$$J_{\text{diffn}} = q D_n \frac{dn}{dx}$$

where D_n is the electron diffusion coefficient, and $\frac{dn}{dx}$ represents the concentration gradient of electrons in the n-side.

Similarly, for holes in the p-side, the diffusion current density (J_{diffp}) can be expressed as:

$$J_{\text{diffp}} = -q D_p \frac{dp}{dx}$$

where D_p is the hole diffusion coefficient, and $\frac{dp}{dx}$ denotes the concentration gradient of holes in the p-side.

Assuming that the depletion region is much larger than the diffusion lengths, the concentration gradients can be approximated to zero within the depletion region. Furthermore, considering that the electron and hole currents are equal under forward bias, we can express the total current density (J) as:

$$J = J_{\text{driftn}} + J_{\text{driftp}}$$

Substituting the expressions for the drift currents, we obtain:

$$J = q(n\mu_n + p\mu_p)\frac{\Delta V}{L_n}$$

where L_n denotes the length of the n-side.

The carrier concentrations (n and p) can be related to the doping concentrations (N_D and N_A) and the intrinsic carrier concentration (n_i). Under thermal equilibrium, they can be expressed as:

$$n = n_i^2/p$$

$$p = n_i^2/n$$

Substituting these relations into the equation for the total current density, we have:

$$J = q(n_i^2/p\mu_n + n_i^2/n\mu_p)\frac{\Delta V}{L_n}$$

Simplifying the equation further, we obtain:

$$J = J_0\left(e^{(q\Delta V)/(nk_B T)} - 1\right)$$

where J_0 is known as the saturation current density and is given by:

$$J_0 = qn_i^2\left(\frac{\mu_n}{p} + \frac{\mu_p}{n}\right)$$

k_B represents Boltzmann's constant, and T denotes the temperature.

This equation, known as the Shockley equation, relates the current density (J) to the voltage bias (ΔV), temperature (T), and other intrinsic properties of the semiconductor material, such as the intrinsic carrier concentration (n_i) and carrier mobilities (μ_n and μ_p).

Implications and Applications

The Shockley equation provides important insights into the behavior of pn junctions and has numerous practical applications. Some key implications and applications of the equation include:

1. **Forward and Reverse Bias**: The Shockley equation helps us understand the behavior of a pn junction under both forward

and reverse bias conditions. Under forward bias, the exponential term dominates, resulting in a significant current flow across the junction. In contrast, under reverse bias, the exponential term tends to zero, leading to negligible current flow.

2. **Current-voltage Characteristics**: The Shockley equation allows us to analyze the current-voltage (IV) characteristics of a pn junction. By plotting the current density against the voltage bias, the IV characteristics can be obtained. The IV curves provide valuable information about the junction's behavior and allow for the estimation of important parameters such as the ideality factor and series resistance.

3. **Diode Modeling**: The Shockley equation forms the basis for modeling diodes, which are essential components in electronic circuits. By incorporating the Shockley equation into circuit simulations, the behavior of diodes under different operating conditions can be accurately predicted, enabling the design and optimization of electronic systems.

4. **Device Efficiency and Performance**: The Shockley equation plays a crucial role in the analysis and optimization of various semiconductor devices, including solar cells and light-emitting diodes (LEDs). Understanding the current-voltage characteristics provided by the equation allows for the improvement of device efficiency and performance through the design of appropriate doping profiles and material properties.

In summary, the Shockley equation forms the foundation for understanding the current-voltage characteristics of pn junctions and has numerous applications in semiconductor device modeling and analysis. Its derivation and implications provide valuable insights into the behavior of junctions under different biasing conditions, aiding in the design and optimization of a wide range of electronic and optoelectronic devices.

Python Code Snippet

Below is a Python code snippet that implements the Shockley equation for a pn junction, calculating the current density based on applied voltage, intrinsic carrier concentration, and carrier mobilities.

```
def shockley_equation(voltage, n_i, mu_n, mu_p, T, q):
    '''
    Calculate the current density for a pn junction based on the
    ↪ Shockley equation.
```

```
:param voltage: Voltage applied across the junction (in volts).
:param n_i: Intrinsic carrier concentration (in m^-3).
:param mu_n: Electron mobility (in m^2/(V·s)).
:param mu_p: Hole mobility (in m^2/(V·s)).
:param T: Temperature (in Kelvin).
:param q: Elementary charge (in coulombs).
:return: Current density (in A/m^2).
'''
    k_B = 1.380649e-23  # Boltzmann's constant in J/K
    # Calculate the saturation current density (J0)
    J0 = q * n_i**2 * (mu_n + mu_p) / (n_i + n_i)
    # Apply the Shockley equation
    J = J0 * (np.exp(q * voltage / (k_B * T)) - 1)
    return J


# Constants
q = 1.602176634e-19  # Elementary charge in coulombs
T = 300  # Temperature in Kelvin
n_i = 1.5e10  # Intrinsic carrier concentration in m^-3 for silicon
mu_n = 0.145  # Electron mobility in m^2/(V·s)
mu_p = 0.048  # Hole mobility in m^2/(V·s)

# Voltage across the junction
voltage = 0.7  # Forward bias voltage in volts

# Calculate current density
current_density = shockley_equation(voltage, n_i, mu_n, mu_p, T, q)

# Output the results
print("Current Density:", current_density, "A/m^2")
```

This code defines a function:

- `shockley_equation` calculates the current density across a pn junction given the voltage applied, intrinsic carrier concentration, carrier mobilities, temperature, and elementary charge.

The provided example uses typical values for silicon at room temperature to compute the current density when a forward bias of 0.7 volts is applied across the junction, and then prints the result.

Chapter 18

MOS Capacitor

The MOS (Metal-Oxide-Semiconductor) capacitor is a fundamental structure in semiconductor device physics, widely used in integrated circuits for various applications. In this chapter, we delve into the capacitance-voltage characteristics and surface potential of the MOS capacitor. We explore the underlying mathematics and provide expert insights into the behavior of this important device.

Capacitance-Voltage Characteristics

The capacitance-voltage (CV) characteristics of a MOS capacitor describe the relationship between the voltage applied across the device and the resulting capacitance. It provides valuable information about the charge distribution, carrier concentrations, and energy bands in the device structure.

1 Basic Theory

The basic theory behind the CV characteristics of a MOS capacitor is rooted in the fundamental principles of charge storage and electrostatics. A MOS capacitor consists of a metal gate electrode separated from a semiconductor substrate by a thin insulating layer, typically made of silicon dioxide.

When a voltage bias is applied to the device, the electric field generated between the gate and the substrate modifies the charge distribution in the semiconductor. This results in the formation of depletion regions near the oxide-substrate interface.

The capacitance of the MOS capacitor is given by the equation:

$$C = \frac{dQ}{dV} \qquad (18.1)$$

Where C is the capacitance, dQ is the change in charge, and dV is the change in voltage.

The charge dQ can be expressed as the sum of the charge in the depletion region (dQ_{dep}), the charge in the inversion region (dQ_{inv}), and the charge in the oxide layer (dQ_{ox}).

2 Depletion Region Capacitance

The depletion region capacitance (C_{dep}) is determined by the charge variation in the depletion region. Under reverse bias conditions, the width of the depletion region increases, leading to an increase in the depletion region capacitance.

The depletion region capacitance can be given by:

$$C_{\text{dep}} = \sqrt{\frac{2\epsilon_s \epsilon_0}{q}\left(\frac{N_a + N_d}{N_a N_d}\right)(\Phi - V)} \qquad (18.2)$$

Where ϵ_s is the permittivity of the semiconductor, ϵ_0 is the permittivity of free space, N_a and N_d are the acceptor and donor concentrations in the semiconductor, Φ is the potential barrier at the oxide-semiconductor interface, and V is the voltage applied to the MOS capacitor.

3 Inversion Region Capacitance

As the voltage bias increases and becomes positive, a point is reached where the surface potential is sufficient to create an inversion layer near the oxide-semiconductor interface. This inversion layer is formed by minority carriers in the semiconductor, creating an additional charge that affects the capacitance of the MOS capacitor.

The inversion region capacitance (C_{inv}) can be expressed as:

$$C_{\text{inv}} = \frac{\epsilon_s \epsilon_0}{d_{\text{inv}}} \qquad (18.3)$$

Where ϵ_s is the permittivity of the semiconductor, ϵ_0 is the permittivity of free space, and d_{inv} is the thickness of the inversion region.

4 Oxide Capacitance

The oxide capacitance (C_{ox}) arises from the charge storage in the insulating oxide layer. It can be given by:

$$C_{ox} = \frac{\epsilon_{ox}\epsilon_0}{t_{ox}} \tag{18.4}$$

Where ϵ_{ox} is the permittivity of the oxide layer and t_{ox} is the thickness of the oxide layer.

5 Total Capacitance

The total capacitance of the MOS capacitor is obtained by summing the depletion region capacitance (C_{dep}), inversion region capacitance (C_{inv}), and oxide capacitance (C_{ox}). Therefore, the total capacitance (C) can be expressed as:

$$C = C_{dep} + C_{inv} + C_{ox} \tag{18.5}$$

Surface Potential

The surface potential of a MOS capacitor refers to the electrostatic potential near the oxide-semiconductor interface. It is a key parameter in MOS device modeling and is closely related to the carrier concentration and band bending in the semiconductor.

1 Basic Theory

The surface potential (ψ_s) is related to the charge density in the semiconductor substrate. Under thermal equilibrium, the surface potential can be obtained by solving Poisson's equation:

$$\frac{d^2\psi_s}{dx^2} = -\frac{q}{\epsilon_s}(N_d - N_a) \tag{18.6}$$

Where x is the distance from the oxide-semiconductor interface, q is the elementary charge, ϵ_s is the permittivity of the semiconductor, N_d is the donor concentration, and N_a is the acceptor concentration.

2 Flatband Voltage

The flatband voltage (V_{FB}) of a MOS capacitor refers to the voltage at which the surface potential becomes constant throughout the semiconductor substrate. It can be determined by solving Poisson's equation while considering the absence of any space charge in the semiconductor.

At the flatband voltage, the net charge density in the semiconductor is zero. Therefore, Poisson's equation can be simplified to:

$$\frac{d^2 \psi_s}{dx^2} = 0 \tag{18.7}$$

Solving this equation yields a constant surface potential value equal to the flatband voltage.

3 Threshold Voltage

The threshold voltage (V_{th}) of a MOS capacitor represents the voltage at which the inversion layer starts to form at the oxide-semiconductor interface. It is a critical parameter for MOS device operation and is affected by various factors such as doping concentration, oxide thickness, and temperature.

The threshold voltage can be calculated by considering the charge in the inversion layer, which is related to the surface potential. It can be expressed as:

$$V_{th} = V_{FB} + 2\Phi_F \tag{18.8}$$

Where V_{FB} is the flatband voltage and Φ_F is the Fermi potential, representing the difference between the energy levels of the semiconductor and the metal gate electrode.

4 Subthreshold Swing

The subthreshold swing (S) is a parameter that characterizes the variation in the surface potential with respect to the applied voltage. It quantifies the sensitivity of the transistor to changes in gate voltage in the subthreshold region, where the device operates with low current levels.

The subthreshold swing can be given by:

$$S = \frac{dV}{d(\log(I))} \tag{18.9}$$

Where V is the applied voltage and I is the current.

Conclusion

In this chapter, we explored the capacitance-voltage characteristics and surface potential of the MOS capacitor. We derived the expressions for the depletion region capacitance, inversion region capacitance, and oxide capacitance, providing a deep understanding of the underlying mathematics. Additionally, we discussed the surface potential and its relationship with the charge density in the semiconductor substrate. The insights gained from this chapter form a foundation for the analysis and design of MOS devices, enabling the development of advanced integrated circuits.

Python Code Snippet

Below is a Python code snippet that implements the key equations and algorithms discussed in the chapter regarding MOS capacitor characteristics, including capacitance calculations and surface potential analysis.

```python
import math

def calculate_depletion_region_capacitance(N_a, N_d, V, Phi,
    epsilon_s, epsilon_0):
    '''
    Calculate depletion region capacitance of a MOS capacitor.
    :param N_a: Acceptor concentration in m^-3.
    :param N_d: Donor concentration in m^-3.
    :param V: Applied voltage in volts.
    :param Phi: Potential barrier at the oxide-semiconductor
        interface in volts.
    :param epsilon_s: Permittivity of the semiconductor in F/m.
    :param epsilon_0: Permittivity of free space in F/m.
    :return: Depletion region capacitance in F.
    '''
    C_dep = math.sqrt((2 * epsilon_s * epsilon_0 / 1.6e-19) * ((N_a
        + N_d) / (N_a * N_d)) * (Phi - V))
    return C_dep


def calculate_inversion_region_capacitance(epsilon_s, epsilon_0,
    d_inv):
    '''
    Calculate inversion region capacitance of a MOS capacitor.
```

```python
    :param epsilon_s: Permittivity of the semiconductor in F/m.
    :param epsilon_0: Permittivity of free space in F/m.
    :param d_inv: Thickness of the inversion region in meters.
    :return: Inversion region capacitance in F.
    '''
    C_inv = epsilon_s * epsilon_0 / d_inv
    return C_inv


def calculate_oxide_capacitance(epsilon_ox, t_ox):
    '''
    Calculate the oxide capacitance of a MOS capacitor.
    :param epsilon_ox: Permittivity of the oxide layer in F/m.
    :param t_ox: Thickness of the oxide layer in meters.
    :return: Oxide capacitance in F.
    '''
    C_ox = epsilon_ox * 8.854187e-12 / t_ox  # epsilon_0 value in
    ↪ F/m
    return C_ox


def calculate_total_capacitance(C_dep, C_inv, C_ox):
    '''
    Calculate the total capacitance of a MOS capacitor.
    :param C_dep: Depletion region capacitance in F.
    :param C_inv: Inversion region capacitance in F.
    :param C_ox: Oxide capacitance in F.
    :return: Total capacitance in F.
    '''
    return C_dep + C_inv + C_ox


def calculate_surface_potential(N_d, N_a, epsilon_s, V):
    '''
    Calculate surface potential using Poisson's equation.
    :param N_d: Donor concentration in m^-3.
    :param N_a: Acceptor concentration in m^-3.
    :param epsilon_s: Permittivity of the semiconductor in F/m.
    :param V: Voltage applied to the MOS capacitor in volts.
    :return: Surface potential in volts.
    '''
    return (1/(epsilon_s/(1.6e-19))) * (N_d - N_a) * V  # Simplified
    ↪ from Poisson's equation


# Example parameter values for MOS capacitor
N_a = 1e24  # Acceptor concentration in m^-3
N_d = 1e24  # Donor concentration in m^-3
V = 3.0  # Applied voltage in volts
Phi = 0.7  # Potential barrier in volts
epsilon_s = 11.7 * 8.854187e-12  # Permittivity of silicon in F/m
epsilon_ox = 3.45 * 8.854187e-12  # Permittivity of silicon dioxide
↪ in F/m
```

```python
t_ox = 1e-9  # Oxide thickness in meters
d_inv = 1e-8  # Inversion layer thickness in meters

# Calculations
C_dep = calculate_depletion_region_capacitance(N_a, N_d, V, Phi,
    epsilon_s, 8.854187e-12)
C_inv = calculate_inversion_region_capacitance(epsilon_s,
    8.854187e-12, d_inv)
C_ox = calculate_oxide_capacitance(epsilon_ox, t_ox)
C_total = calculate_total_capacitance(C_dep, C_inv, C_ox)
surface_potential = calculate_surface_potential(N_d, N_a, epsilon_s,
    V)

# Output results
print("Depletion Region Capacitance:", C_dep, "F")
print("Inversion Region Capacitance:", C_inv, "F")
print("Oxide Capacitance:", C_ox, "F")
print("Total Capacitance:", C_total, "F")
print("Surface Potential:", surface_potential, "V")
```

This code defines several functions:

- `calculate_depletion_region_capacitance` computes the capacitance of the depletion region based on semiconductor properties and voltage.
- `calculate_inversion_region_capacitance` calculates the capacitance of the inversion region.
- `calculate_oxide_capacitance` derives the oxide capacitance using the permittivity and thickness of the oxide layer.
- `calculate_total_capacitance` provides the total capacitance of the MOS capacitor by summing its components.
- `calculate_surface_potential` estimates the surface potential of the MOS capacitor using Poisson's equation.

The example provided in the code calculates the various capacitances and surface potential for a MOS capacitor with given parameter values, then prints out the results.

Chapter 19

Band Bending

The phenomenon of band bending plays a crucial role in understanding the behavior of interfaces and devices in semiconductors. It arises due to the variation in the energy band structure near the interface between different materials or in the presence of external electric fields. In this chapter, we delve into the mathematics and physical insights behind band bending, shedding light on its implications for semiconductor physics.

Basic Theory and Governing Equations

The concept of band bending stems from the fundamental principles of electrostatics and quantum mechanics. It describes the bending or tilting of energy bands near an interface or within a device due to charge imbalances and the resulting electrostatic potential distribution.

Band bending is typically quantified by considering the energy level shift, i.e., the difference in the energy position of the bands with respect to a reference level, such as the vacuum level or a fixed energy reference within the material.

1 Poisson's Equation

In order to analyze band bending, we employ Poisson's equation, derived from Maxwell's equations and Gauss's law. Poisson's equation relates the electric field (E) to the charge density (ρ) as follows:

$$\nabla^2 \varphi = -\frac{\rho}{\epsilon} \tag{19.1}$$

where φ denotes the electric potential, ∇^2 represents the Laplacian operator, ρ is the charge density, and ϵ is the permittivity of the material.

2 Space Charge Region

The key factor underlying band bending is the presence of a space charge region (SCR) near the interface or in the vicinity of a p-n junction. Due to the presence of ionized impurities or immobile charges, the space charge region contains a net electric charge. This charge density, attributed to ionized dopant atoms or trapped charges, induces an electric field across the SCR.

3 Governing Equations

To examine the band bending in the SCR, we consider both the charge continuity equation and the Poisson's equation. The charge continuity equation describes the balance between charge recombination and generation, while Poisson's equation describes the charge distribution and potential profile. The coupled equations can be expressed as follows:

$$\frac{\partial p}{\partial t} - \nabla \cdot (D_p \nabla p) + G - R = 0 \qquad (19.2)$$

for the equation governing holes, and

$$\frac{\partial n}{\partial t} - \nabla \cdot (D_n \nabla n) + G - R = 0 \qquad (19.3)$$

for the equation governing electrons. Here, p and n represent the hole and electron densities, respectively, D_p and D_n are the diffusion constants, G and R denote the generation and recombination terms, and t represents time.

4 Surface Potential

The surface potential, denoted by Ψ_s, corresponds to the change in the electrostatic potential across the space charge region. It quantifies the difference in potential between the interface and a reference position.

The relationship between the surface potential and the charge distribution can be derived by integrating the electric field along a specific path within the semiconductor. The resulting equation, referred to as the surface potential equation, relates the surface potential to the charge density and other parameters.

5 Intrinsic Semiconductor

In the case of an intrinsic semiconductor, where the concentrations of dopants are negligibly small, the charge continuity equations can be simplified. For an intrinsic semiconductor without any externally applied biases, the equations governing the hole and electron densities reduce to:

$$\nabla \cdot (D_p \nabla p) = R_p \quad (19.4)$$

and

$$\nabla \cdot (D_n \nabla n) = R_n \quad (19.5)$$

where R_p and R_n represent the recombination terms for holes and electrons, respectively.

Metal-Semiconductor Junctions

The behavior of band bending is particularly pronounced at metal-semiconductor junctions, where the different work functions of the materials lead to the formation of a built-in potential. The charge redistribution across the SCR creates a potential energy barrier, which affects the carrier transport and device characteristics.

1 Schottky Barrier

At a metal-semiconductor junction, a Schottky barrier is formed due to the difference in the work functions of the two materials. This barrier, characterized by band bending, inhibits the flow of majority carriers across the junction and influences the contact properties.

The height of the Schottky barrier, denoted by Φ_B, can be determined by the difference in the Fermi levels of the semiconductor and the metal. It affects the energy band diagram and determines the built-in potential across the junction.

2 PN Junctions

In the case of a p-n junction, band bending arises due to the diffusion of majority carriers from regions of high concentration to low concentration. The resulting space charge region includes both ionized donors and acceptors, leading to a potential barrier.

The depletion region of a p-n junction, characterized by the presence of band bending, plays a crucial role in the rectifying behavior and diode characteristics. The width of the depletion region strongly influences the junction capacitance and breakdown voltage.

Conclusion

In this chapter, we explored the phenomenon of band bending and its significance in interface and device physics. By examining the mathematical equations governing charge distribution and potential profiles, we gained insights into the intricate behavior of energy bands near interfaces. We specifically focused on metal-semiconductor junctions, including Schottky barriers and p-n junctions, and discussed the implications of band bending for carrier transport and device performance.

Python Code Snippet

Below is a Python code snippet that implements the governing equations and concepts discussed in the chapter on band bending, including Poisson's equation, charge continuity equations, and the calculation of the Schottky barrier height.

```python
import numpy as np
import matplotlib.pyplot as plt

def poisson_equation(charge_density, epsilon, x):
    '''
    Solve Poisson's equation to determine electric potential.
    :param charge_density: Charge density distribution (C/m^3).
    :param epsilon: Permittivity of the material (F/m).
    :param x: Spatial coordinate (m).
    :return: Electric potential (V).
    '''
    dx = x[1] - x[0]   # Assume uniform spacing
    potential = np.zeros_like(charge_density)
    # Solve Poisson's equation using finite difference method
```

```python
    for i in range(1, len(potential)-1):
        potential[i] = potential[i-1] + (dx**2 / epsilon) *
        ↪ charge_density[i-1]
    return potential

def schottky_barrier_height(work_function_metal,
↪ electron_affinity_semiconductor):
    '''
    Calculate the height of the Schottky barrier.
    :param work_function_metal: Work function of the metal (eV).
    :param electron_affinity_semiconductor: Electron affinity of the
    ↪ semiconductor (eV).
    :return: Schottky barrier height (eV).
    '''
    return work_function_metal - electron_affinity_semiconductor

def charge_continuity(hole_density, electron_density,
↪ generation_rate, recombination_rate, D_p, D_n):
    '''
    Solve charge continuity equations for holes and electrons.
    :param hole_density: Density of holes (m^-3).
    :param electron_density: Density of electrons (m^-3).
    :param generation_rate: Rate of generation (m^-3s^-1).
    :param recombination_rate: Rate of recombination (m^-3s^-1).
    :param D_p: Diffusion constant for holes (m^2/s).
    :param D_n: Diffusion constant for electrons (m^2/s).
    :return: New hole and electron densities.
    '''
    # Apply continuity for charge carriers (assumed simple model)
    hole_density_new = hole_density + generation_rate -
    ↪ recombination_rate
    electron_density_new = electron_density + generation_rate -
    ↪ recombination_rate
    return hole_density_new, electron_density_new

# Parameters for the calculations
epsilon = 8.854E-12  # Permittivity of free space in F/m
work_function_metal = 4.5  # Work function for the metal in eV
electron_affinity_semiconductor = 4.0  # Electron affinity of
↪ semiconductor in eV
length = 1e-6  # Length of the semiconductor in meters
x = np.linspace(0, length, 100)  # Spatial coordinates

# Sample charge density (for testing)
charge_density = np.linspace(1E20, 1E18, x.size)  # Charge density
↪ in C/m^3

# Calculate electric potential
potential = poisson_equation(charge_density, epsilon, x)

# Calculate Schottky barrier height
barrier_height = schottky_barrier_height(work_function_metal,
↪ electron_affinity_semiconductor)
```

```python
# Initial parameters for charge continuity
hole_density = 1E15    # Initial hole density in m^-3
electron_density = 1E16    # Initial electron density in m^-3
generation_rate = 1E20    # Generation rate in m^-3s^-1
recombination_rate = 5E19    # Recombination rate in m^-3s^-1
D_p = 1E-5    # Diffusion constant for holes in m^2/s
D_n = 1E-5    # Diffusion constant for electrons in m^2/s

# Update charge densities based on continuity
new_hole_density, new_electron_density =
 ↪ charge_continuity(hole_density, electron_density,
 ↪ generation_rate, recombination_rate, D_p, D_n)

# Output results
print("Electric Potential at interface (V):", potential[-1])
print("Schottky Barrier Height (eV):", barrier_height)
print("New Hole Density (m^-3):", new_hole_density)
print("New Electron Density (m^-3):", new_electron_density)

# Plotting electric potential
plt.plot(x, potential)
plt.title('Electric Potential Distribution')
plt.xlabel('Position (m)')
plt.ylabel('Electric Potential (V)')
plt.grid()
plt.show()
```

This code defines three functions:

- `poisson_equation` calculates the electric potential based on the charge density using Poisson's equation.
- `schottky_barrier_height` computes the height of the Schottky barrier based on the work function of the metal and the electron affinity of the semiconductor.
- `charge_continuity` updates the hole and electron densities using a simple model of generation and recombination.

The provided example calculates the electric potential at an interface, the Schottky barrier height, updates carrier densities based on generation and recombination, and plots the electric potential distribution.

Chapter 20

Quantum Tunneling

In this chapter, we explore the phenomenon of quantum tunneling and its impact on semiconductor devices. Quantum tunneling is a quantum mechanical phenomenon in which particles can pass through potential energy barriers, even when their energy is lower than the height of the barrier. This remarkable behavior plays a crucial role in various aspects of semiconductor physics and has significant implications for device design and performance.

Introduction to Quantum Tunneling

Quantum tunneling arises from the wave-particle duality of quantum mechanics. According to the Heisenberg uncertainty principle, there is an inherent uncertainty in the position and momentum of particles. This uncertainty gives rise to the possibility of particles existing in energetically forbidden regions, allowing them to penetrate potential energy barriers.

1 Barrier Penetration Probability

The probability of a particle tunneling through a potential barrier depends on various factors, such as the width and height of the barrier, the energy of the particle, and the mass of the particle. The tunneling probability can be described using the WKB approximation (Wentzel-Kramers-Brillouin approximation) or by solving the Schrödinger equation directly.

Impact on Semiconductor Devices

Quantum tunneling has a profound impact on the behavior and performance of semiconductor devices. Understanding and controlling tunneling effects are crucial for the design and optimization of electronic devices. We examine some of the key aspects of tunneling in semiconductors in the following subsections.

1 Tunneling in Transistors

In metal-oxide-semiconductor field-effect transistors (MOSFETs), the phenomenon of tunneling plays a critical role in the operation of the device. The thin insulating oxide layer acts as a potential barrier, allowing carriers to tunnel from the source to the drain region. This tunneling current, known as the subthreshold leakage current, affects the overall device performance and power consumption.

2 Tunneling in Diodes

Diode devices also exhibit tunneling effects, which can impact their characteristics and performance. In p-n junctions, for example, carriers can tunnel through the depletion region, contributing to the reverse current known as the diode leakage current. Tunneling currents become significant at smaller device dimensions, leading to leakage issues and potential reliability concerns.

3 Quantum Well Structures

Quantum well structures, consisting of thin layers of semiconductors with different bandgap energies, take advantage of quantum tunneling effects. These structures confine carriers in a potential well, enabling quantum tunneling across the potential barriers. Quantum well devices, such as quantum well lasers, exhibit unique properties and enhanced performance due to the quantized energy levels and increased carrier confinement.

4 Tunneling Transistors

Emerging transistor designs, such as tunnel field-effect transistors (TFETs), utilize quantum tunneling as the main mechanism for carrier transport. TFETs exploit the tunneling of carriers through the potential barrier in the source-drain junction, enabling low-power operation and potential improvements in energy efficiency compared to conventional transistor technologies.

Conclusion

In this chapter, we examined the phenomenon of quantum tunneling and its impact on semiconductor devices. Quantum tunneling, a unique quantum mechanical behavior, enables particles to penetrate potential energy barriers, defying classical expectations. In the context of semiconductor devices, quantum tunneling plays a significant role in phenomena such as subthreshold leakage in MOSFETs, diode leakage currents, and the behavior of quantum well structures. The understanding and control of tunneling effects are crucial for optimizing device performance and exploring novel transistor designs.

Python Code Snippet

Below is a Python code snippet that calculates the tunneling probability through a potential barrier and analyzes the impact of quantum tunneling on semiconductor devices.

```python
import numpy as np
import matplotlib.pyplot as plt

def tunneling_probability(V0, E, a, m):
    '''
    Calculate the tunneling probability using the WKB approximation.
    :param V0: Height of the potential barrier in joules.
    :param E: Energy of the particle in joules.
    :param a: Width of the potential barrier in meters.
    :param m: Mass of the particle in kg.
    :return: Tunneling probability (unitless).
    '''
    if E >= V0:
        return 1  # No tunneling if E is greater than or equal to V0
    else:
        kappa = ((2 * m * (V0 - E))**0.5)  # Kappa for exponential
        ↪ decay
        return np.exp(-2 * kappa * a / hbar)

def plot_tunneling_probability(V0, E, a, m, a_range):
    '''
    Plot tunneling probability as a function of barrier width.
```

```
:param V0: Height of the potential barrier in joules.
:param E: Energy of the particle in joules.
:param m: Mass of the particle in kg.
:param a_range: Range of width values (in meters) for the plot.
'''
    probabilities = [tunneling_probability(V0, E, width, m) for
    ↪ width in a_range]
    plt.figure(figsize=(8, 5))
    plt.plot(a_range, probabilities, label='Tunneling Probability',
    ↪ color='blue')
    plt.title('Tunneling Probability vs. Barrier Width')
    plt.xlabel('Barrier Width (m)')
    plt.ylabel('Tunneling Probability (unitless)')
    plt.yscale('log')  # Logarithmic scale for better visualization
    plt.grid(True)
    plt.legend()
    plt.show()

# Constants and Inputs
hbar = 1.054571e-34  # Reduced Planck's constant in J.s
V0 = 1.6e-19  # Height of the potential barrier in joules (example
↪ value)
E = 1.0e-19  # Particle energy in joules (example value)
m = 9.11e-31  # Mass of an electron in kg
a_range = np.linspace(1e-10, 5e-9, 100)  # Barrier width from 0.1 nm
↪ to 5 nm

# Calculate and plot tunneling probability
plot_tunneling_probability(V0, E, a_range[0], m, a_range)
```

This code includes two functions:

- `tunneling_probability` calculates the tunneling probability through a potential barrier using the WKB approximation based on the height, width, and energy of the barrier as well as the mass of the particle.
- `plot_tunneling_probability` generates a graph showing the tunneling probability as a function of the barrier width.

The provided example sets specific values for the potential barrier height, particle energy, and mass, while also creating a range of barrier widths to visualize the relationship between tunneling probability and barrier width. The tunneling probability is then plotted in a logarithmic scale for better visualization.

Chapter 21

Charge Control Model

The charge control model is an essential concept in the field of semiconductor physics, specifically applied to Metal-Oxide-Semiconductor Field-Effect Transistors (MOSFETs). In this chapter, we explore the application of the charge control model in MOSFETs to understand two critical parameters: the threshold voltage and transconductance. By utilizing mathematical equations and analysis, we can gain deeper insights into the behavior and performance of these devices.

Introduction to Charge Control Model

The charge control model is a fundamental framework used to describe the behavior of MOSFETs. It provides insights into the relationship between the charges in the device and the external voltages applied to it. By understanding this relationship, we can analyze important device characteristics, such as the threshold voltage and transconductance.

1 Metal-Oxide-Semiconductor Structure

A MOSFET consists of a Metal-Oxide-Semiconductor (MOS) structure, where a layer of insulating material (usually silicon dioxide, SiO_2) separates the metal gate from the semiconductor region (typically doped silicon). The gate voltage controls the charge distribution in the semiconductor, modulating the device's conductivity.

2 Threshold Voltage

The threshold voltage (V_{th}) is a crucial parameter in MOSFET operation. It defines the gate voltage at which the device transitions from the off-state to the on-state. The charge control model provides a mathematical expression for determining the threshold voltage.

The charge per unit area in the semiconductor, denoted as Q_{inv}, can be expressed as:

$$Q_{inv} = -C_{ox}(V_{gs} - V_{th})$$

where C_{ox} represents the capacitance per unit area of the oxide layer, V_{gs} is the gate-to-source voltage, and V_{th} is the threshold voltage. This equation indicates that the charge in the semiconductor is proportional to the difference between the gate voltage and the threshold voltage.

3 Transconductance

Transconductance (g_m) is another critical parameter in MOSFETs and represents the ratio of the change in the device's drain current (I_d) to the change in the gate-to-source voltage (V_{gs}). It characterizes the device's ability to convert changes in the gate voltage into current variation.

The transconductance can be derived from the charge control model using the following equation:

$$g_m = \frac{dI_d}{dV_{gs}} = \sqrt{2\mu_n C_{ox} I_d}$$

where μ_n represents the electron mobility in the semiconductor and I_d is the drain current. This equation shows that the transconductance is proportional to the square root of the product of the electron mobility, oxide capacitance, and drain current.

Conclusion

In this chapter, we explored the application of the charge control model in MOSFETs, specifically focusing on understanding the threshold voltage and transconductance. The charge control model allows us to analyze the relationship between charges in the device and external voltages, providing valuable insights into device behavior.

By utilizing mathematical equations derived from the charge control model, we can calculate the threshold voltage and understand how it dictates the transition between the off-state and on-state of the MOSFET. Additionally, the transconductance equation enables us to evaluate the device's ability to convert changes in gate voltage into current variations.

Understanding and utilizing the charge control model and its associated equations are vital for designing and optimizing MOSFET devices, leading to advancements in semiconductor technology and integrated circuits.

Python Code Snippet

Below is a Python code snippet that calculates the threshold voltage and transconductance for a MOSFET based on the charge control model.

```python
def calculate_threshold_voltage(V_gs, C_ox):
    '''
    Calculate the threshold voltage for a MOSFET using the charge
    ↪ control model.
    :param V_gs: Gate-to-source voltage in volts.
    :param C_ox: Capacitance per unit area of oxide in F/m^2.
    :return: Threshold voltage in volts.
    '''
    # Assuming V_th is initially set to zero for simplicity; can be
    ↪ updated as necessary
    V_th = 0  # You can modify this for more precise calculations
    Q_inv = -C_ox * (V_gs - V_th)  # Charge per unit area
    return V_th + (Q_inv / -C_ox)  # Derived threshold voltage


def calculate_transconductance(I_d, mu_n, C_ox):
    '''
    Calculate the transconductance for a MOSFET.
    :param I_d: Drain current in amperes.
    :param mu_n: Electron mobility in m^2/Vs.
    :param C_ox: Capacitance per unit area of oxide in F/m^2.
    :return: Transconductance in siemens (S).
    '''
    g_m = (2 * mu_n * C_ox * I_d) ** 0.5  # Transconductance
    ↪ calculation
    return g_m
```

```python
# Inputs for the calculations
V_gs = 5.0      # Gate-to-source voltage in volts
C_ox = 3.45e-3  # Oxide capacitance per unit area in F/m^2
I_d = 0.01      # Drain current in amperes (10 mA)
mu_n = 0.15     # Electron mobility in m^2/Vs

# Calculations
threshold_voltage = calculate_threshold_voltage(V_gs, C_ox)
transconductance = calculate_transconductance(I_d, mu_n, C_ox)

# Output results
print("Threshold Voltage:", threshold_voltage, "V")
print("Transconductance:", transconductance, "S")
```

This code defines two functions:

- `calculate_threshold_voltage` calculates the threshold voltage for a MOSFET based on gate-to-source voltage and oxide capacitance.

- `calculate_transconductance` computes the transconductance based on the drain current, electron mobility, and oxide capacitance.

The provided example calculates the threshold voltage and transconductance for a MOSFET given specific parameters, then prints the results.

Chapter 22

Small-Signal Model

In this chapter, we delve into the analysis of the AC response of semiconductor devices using the small-signal model. The small-signal model is a linear approximation technique widely employed in the field of semiconductor physics and device engineering to analyze the behavior of devices under small perturbations around their quiescent operating points.

Introduction to Small-Signal Model

The small-signal model provides a mathematical framework for characterizing the small-signal behavior of semiconductor devices, facilitating the analysis of amplifiers, filters, and other electronic circuits that rely on the linear response of devices. By linearizing the device equations around their DC operating points, the small-signal model enables the decomposition of complex device behavior into simpler linear components.

1 Linearization

The first step in constructing the small-signal model is to linearize the device equations around their DC operating points. This entails linearizing the nonlinear device equations, such as the diode or transistor equations, by expanding them as Taylor series up to the first-order terms. The DC operating points serve as the reference or quiescent points around which the small-signal behavior will be analyzed.

2 Small-Signal Equivalent Circuit

Once the device equations are linearized, the small-signal equivalent circuit is constructed by replacing each nonlinear device element with its linear small-signal equivalent. This equivalent circuit consists of resistors, capacitors, and controlled current or voltage sources, representing the linearized behavior of the device under small perturbations.

Small-Signal Analysis of Semiconductors

The small-signal model finds extensive application in the analysis of semiconductor devices, such as transistors and amplifiers. It enables the determination of important parameters, including the small-signal gain, input and output impedances, and frequency response.

1 Small-Signal Gain

The small-signal gain (A_v) of a device measures the ratio of the small-signal output voltage to the small-signal input voltage. It characterizes the device's amplification capability within its linear operating range. By utilizing the small-signal model, we can derive mathematical expressions to calculate the small-signal gain of various semiconductor devices.

2 Input and Output Impedances

The input and output impedances are crucial parameters in semiconductor devices, as they dictate the interaction between devices and the connected circuits. The input impedance (Z_{in}) represents the impedance seen by the signal source connected to the device's input, while the output impedance (Z_{out}) represents the impedance seen by the load connected to the device's output. The small-signal model aids in analyzing and determining these impedance values.

3 Frequency Response

The frequency response of semiconductor devices is an essential aspect to consider in their design and analysis. The small-signal model facilitates the calculation of the frequency response characteristics, such as the cut-off frequency, bandwidth, and phase shifts.

By analyzing the small-signal behavior at different frequencies, we gain insights into the device's behavior and performance over the entire frequency range.

Conclusion

In this chapter, we have explored the small-signal model as a critical tool for analyzing the AC response of semiconductor devices. Through linearization and the construction of small-signal equivalent circuits, the small-signal model allows us to examine the device behavior under small perturbations around their quiescent operating points.

The small-signal model provides valuable insights into the amplification capability, input and output impedance, and frequency response of semiconductor devices. By leveraging mathematical techniques and linear approximations, the small-signal model simplifies the analysis of complex device behavior and aids in the design and optimization of electronic circuits.

Python Code Snippet

Below is a Python code snippet that implements important equations and algorithms related to the small-signal model, enabling users to analyze the small-signal gain, input/output impedances, and frequency response of semiconductor devices.

```python
import numpy as np

def calculate_small_signal_gain(V_out, V_in):
    '''
    Calculate the small-signal gain of a device.
    :param V_out: Small-signal output voltage.
    :param V_in: Small-signal input voltage.
    :return: Small-signal gain (A_v).
    '''
    return V_out / V_in


def calculate_input_impedance(R1, R2):
    '''
    Calculate the input impedance of a voltage divider circuit.
    :param R1: Resistance R1 in ohms.
    :param R2: Resistance R2 in ohms.
    :return: Input impedance in ohms.
```

```python
    '''
    return (R1 * R2) / (R1 + R2)

def calculate_output_impedance(R3):
    '''
    Calculate the output impedance of a circuit connected to a load.
    :param R3: Resistance R3 in ohms (output resistance).
    :return: Output impedance in ohms.
    '''
    return R3

def calculate_frequency_response(R, C):
    '''
    Calculate the cut-off frequency of an RC circuit.
    :param R: Resistance in ohms.
    :param C: Capacitance in farads.
    :return: Cut-off frequency in hertz (Hz).
    '''
    return 1 / (2 * np.pi * R * C)


# Example parameters
V_out = 2.0   # Small-signal output voltage in volts
V_in = 1.0    # Small-signal input voltage in volts
R1 = 1000     # Resistance R1 in ohms
R2 = 2000     # Resistance R2 in ohms
R3 = 3000     # Output resistance in ohms
C = 1e-6      # Capacitance in farads (1 µF)

# Calculations
small_signal_gain = calculate_small_signal_gain(V_out, V_in)
input_impedance = calculate_input_impedance(R1, R2)
output_impedance = calculate_output_impedance(R3)
cut_off_frequency = calculate_frequency_response(R3, C)

# Output results
print("Small-Signal Gain (A_v):", small_signal_gain)
print("Input Impedance (Z_in):", input_impedance, "ohms")
print("Output Impedance (Z_out):", output_impedance, "ohms")
print("Cut-off Frequency:", cut_off_frequency, "Hz")
```

This code defines four functions:

- `calculate_small_signal_gain` computes the small-signal gain based on output and input voltages.
- `calculate_input_impedance` determines the input impedance of a voltage divider circuit.
- `calculate_output_impedance` assesses the output impedance of

a circuit based on the connected load.
- `calculate_frequency_response` calculates the cut-off frequency of an RC circuit.

The provided example uses specified parameters to calculate the small-signal gain, input and output impedances, and cut-off frequency, then prints the results.

Chapter 23

Boltzmann Transport Equation

In this chapter, we will discuss the Boltzmann Transport Equation, which serves as a crucial framework for modeling electronic transport in semiconductors. The Boltzmann Transport Equation provides a comprehensive mathematical description of the transport phenomena, enabling the understanding and analysis of carrier dynamics, conductivity, and mobility in semiconductor devices.

Introduction to the Boltzmann Transport Equation

The Boltzmann Transport Equation (BTE) is a partial differential equation that describes the time evolution of electron or hole distribution in momentum and position space. It takes into account various scattering mechanisms and interactions, such as phonon scattering, impurity scattering, and carrier-carrier scattering. By solving the BTE, we can obtain valuable insights into the behavior of carriers in semiconductors under the influence of external fields and local scatterers.

The general form of the BTE is as follows:

$$\frac{\partial f}{\partial t} + \mathbf{v} \cdot \nabla_{\mathbf{r}} f + \mathbf{F} \cdot \nabla_{\mathbf{k}} f = \left(\frac{\partial f}{\partial t}\right)_{\text{coll}} \qquad (23.1)$$

where $f(\mathbf{k}, \mathbf{r}, t)$ is the carrier distribution function, $\mathbf{v}$ is the carrier velocity, **F** is the external force acting on the carriers, and $\left(\frac{\partial f}{\partial t}\right)_{coll}$ represents the collision integral term accounting for scattering events.

1 Distribution Function

The distribution function $f(\mathbf{k}, \mathbf{r}, t)$ describes the probability density of finding a carrier in a specific quantum state characterized by the wave vector **k**, position **r**, and time t. It provides information about the occupation of energy levels and quantum states available to carriers in the material.

2 Carrier Velocity

The carrier velocity **v** represents the rate of change of carrier position with respect to time. In the presence of an external force **F**, carriers experience an acceleration that affects their velocity. The carrier velocity is related to the carrier energy and effective mass through the equation:

$$\mathbf{v} = \frac{1}{\hbar} \nabla_{\mathbf{k}} E(\mathbf{k}) \tag{23.2}$$

where $\hbar$ is the reduced Planck's constant and $E(\mathbf{k})$ is the energy dispersion relation.

3 Collision Term

The collision integral term $\left(\frac{\partial f}{\partial t}\right)_{coll}$ accounts for the scattering events that occur due to various mechanisms, such as lattice vibrations (phonons), impurities, and carrier-carrier interactions. The collision term captures the change in carrier distribution function due to scattering and enables the study of carrier relaxation and transport dynamics in semiconductors.

Solution Techniques for the Boltzmann Transport Equation

Solving the Boltzmann Transport Equation is a challenging task due to its complex nature and the inclusion of scattering terms.

However, several solution techniques have been developed to approximate and solve the BTE, allowing for the analysis of carrier transport in semiconductors.

1 Relaxation Time Approximation

One commonly used approximation technique is the relaxation time approximation, which assumes that the relaxation time between scattering events is much shorter than the time scale over which the external forces change. This approximation simplifies the BTE by neglecting the time derivatives and leads to the following form of the BTE:

$$\mathbf{v} \cdot \nabla_{\mathbf{r}} f + \mathbf{F} \cdot \nabla_{\mathbf{k}} f = -\frac{f - f_0}{\tau} \qquad (23.3)$$

where f_0 is the equilibrium distribution function and τ is the relaxation time.

2 Monte Carlo Simulation

Monte Carlo simulation is another powerful technique to solve the Boltzmann Transport Equation. In this method, the trajectory of individual carriers is randomly simulated, taking into account scattering events and external forces. By statistically averaging over a large number of carrier trajectories, we can obtain an approximate solution to the BTE. Monte Carlo simulation allows for the investigation of various scattering mechanisms and provides insights into carrier transport and mobility in semiconductors.

3 Numerical Methods

Numerical methods, such as finite difference methods, finite element methods, and Monte Carlo methods, are widely employed to solve the Boltzmann Transport Equation numerically. These methods discretize the position and momentum space, allowing for the approximation of the partial differential equation as a system of algebraic equations. By solving this system iteratively, we can obtain the carrier distribution function and study the transport properties in semiconductors.

Applications of the Boltzmann Transport Equation

The Boltzmann Transport Equation has numerous applications in the field of semiconductor physics and device engineering. It provides a solid theoretical framework for understanding and modeling electronic transport in a variety of semiconductor devices, such as transistors, diodes, and solar cells. The BTE allows for the calculation of important transport parameters, including conductivity, mobility, and carrier lifetime, enabling the design and optimization of semiconductor devices with enhanced performance.

Conclusion

In this chapter, we have explored the Boltzmann Transport Equation as a fundamental framework for modeling electronic transport in semiconductors. By understanding the behavior and dynamics of carriers in semiconductors, we can better comprehend the operation of semiconductor devices and optimize their performance. The Boltzmann Transport Equation serves as a powerful tool in semiconductor physics, providing insights into carrier distribution, conductivity, mobility, and scattering mechanisms. The solution techniques discussed enable the approximation and numerical solution of the BTE to study carrier transport in semiconductors.

Python Code Snippet

Below is a Python code snippet that implements key equations and algorithms discussed in the chapter related to the Boltzmann Transport Equation.

```python
import numpy as np
import matplotlib.pyplot as plt

def boltzmann_transport_equation(f0, force, kx, tau, v):
    """
    Solve the Boltzmann Transport Equation using the relaxation time
    ↪ approximation.
    :param f0: Equilibrium distribution function (numpy array).
    :param force: External force acting on the carriers (numpy
    ↪ array).
    :param kx: Wave vector in the x-direction (numpy array).
```

```python
    :param tau: Relaxation time in seconds.
    :param v: Carrier velocity (numpy array).
    :return: Carrier distribution function (numpy array).
    """
    df_dt = - (f0 - f0.mean()) / tau  # Change in distribution
    ↪  function
    return f0 + df_dt + force * np.gradient(f0, kx)

def calculate_carrier_mobility(sigma, n, q):
    """
    Calculate the carrier mobility.
    :param sigma: Conductivity of the material in S/m.
    :param n: Carrier concentration in m^-3.
    :param q: Charge of the carrier in coulombs.
    :return: Carrier mobility in m^2/(V s).
    """
    return sigma / (n * q)

def calculate_conductivity(mobility, n, q):
    """
    Calculate the conductivity given mobility, carrier
    ↪  concentration, and charge.
    :param mobility: Carrier mobility in m^2/(V s).
    :param n: Carrier concentration in m^-3.
    :param q: Charge of the carrier in coulombs.
    :return: Conductivity in S/m.
    """
    return n * q * mobility

# Parameters for the calculations
num_points = 100
kx = np.linspace(-1e9, 1e9, num_points)  # Wave vector
f0 = np.exp(-kx**2)  # Initial Gaussian distribution
tau = 1e-14  # Relaxation time in seconds
external_force = 1e-3  # External force in N/C
q = 1.6e-19  # Charge of an electron in coulombs
n = 1e25  # Carrier concentration in m^-3
sigma = 0.05  # Conductivity in S/m

# Compute the carrier distribution function
v = kx / np.sqrt(2 * np.pi)  # Assuming some relationship for
↪  velocity
f = boltzmann_transport_equation(f0, external_force, kx, tau, v)

# Calculate mobility and conductivity
mobility = calculate_carrier_mobility(sigma, n, q)
conductivity = calculate_conductivity(mobility, n, q)

# Output results
print("Carrier Mobility:", mobility * 1e6, "µm²/(V s)")
print("Conductivity:", conductivity, "S/m")

# Plotting
```

```
plt.figure(figsize=(10, 6))
plt.plot(kx, f, label='Carrier Distribution Function after
↪ Relaxation')
plt.xlabel('Wave Vector (kx)')
plt.ylabel('Distribution Function f(k)')
plt.title('Boltzmann Transport Equation - Relaxation Time
↪ Approximation')
plt.legend()
plt.grid()
plt.show()
```

This code includes the following functions and computations:

- `boltzmann_transport_equation` solves the Boltzmann Transport Equation using the relaxation time approximation, updating the carrier distribution function based on external force and relaxation effects.
- `calculate_carrier_mobility` computes the carrier mobility from the conductivity, carrier concentration, and charge of the carrier.
- `calculate_conductivity` determines the conductivity based on the calculated mobility, carrier concentration, and charge.

The example provided computes the carrier distribution function, mobility, and conductivity while generating a plot of the carrier distribution function after relaxation, demonstrating the effects of scattering and external forces on carrier dynamics in semiconductors.

Chapter 24

Heterostructures

In this chapter, we delve into the fascinating world of heterostructures, where different materials with distinct energy band structures are combined to create a new material system. Heterostructures play a crucial role in modern semiconductor devices, as they enable precise control over the energy band alignment and transport properties. We will explore the concept of energy band alignment in heterostructures and its significance in determining device performance.

Energy Band Alignment in Heterostructures

In heterostructures, different materials are stacked together, resulting in interfaces between two or more materials. These interfaces give rise to a unique energy band alignment, which is crucial for carrier transport and device operation. The energy band alignment at the interfaces is determined by the properties of each material, such as the bandgap, band-edge positions, and band bending effects.

The energy band alignment in a heterostructure can be understood using the concept of band offsets. A band offset describes the difference in energy between the conduction band minimum (CBM) or valence band maximum (VBM) of one material and the corresponding energy level in the adjacent material. There are two types of band offsets: type I and type II.

1 Type I Band Alignment

In a type I band alignment, the CBM (or VBM) of one material lies above (or below) the CBM (or VBM) of the adjacent material, resulting in a staggered energy band diagram. This alignment creates a potential barrier or well at the interface, influencing carrier transport across the heterojunction. Type I band alignment is commonly observed in materials with similar electronic properties, such as III-V compound semiconductors.

The valence band offset (ΔE_v) represents the energy difference between the VBM levels of the two materials, while the conduction band offset (ΔE_c) represents the energy difference between the CBM levels. The band offsets can be expressed as:

$$\Delta E_v = E_{v1} - E_{v2} \tag{24.1}$$

$$\Delta E_c = E_{c1} - E_{c2} \tag{24.2}$$

where E_{v1} and E_{v2} are the valence band energies, and E_{c1} and E_{c2} are the conduction band energies of the two materials, respectively.

2 Type II Band Alignment

In a type II band alignment, the CBM (or VBM) of one material lies below (or above) the VBM (or CBM) of the adjacent material, resulting in an overlapping energy band diagram. This alignment creates a staggered band edge, leading to a spatially separated electron and hole distribution in the heterostructure. Type II band alignment often occurs in heterostructures involving different materials with disparate electronic properties, such as metal-semiconductor junctions or heterojunctions between materials with different bandgaps.

For a type II band alignment, the band offsets can be defined as:

$$\Delta E_v = E_{c1} - E_{v2} \tag{24.3}$$

$$\Delta E_c = E_{v1} - E_{c2} \tag{24.4}$$

In type II heterostructures, carriers (electrons or holes) are confined to one material due to the energy band misalignment, resulting in unique transport properties and potential applications in devices such as photodetectors and lasers.

Transport Properties in Heterostructures

Heterostructures offer advantageous transport properties by virtue of their tailored band structures. The band offsets at the interfaces influence carrier confinement, tunneling, and recombination processes, enabling the design of novel devices with enhanced performance. The transport properties in heterostructures are strongly dependent on the band alignment and the electronic properties of the constituent materials.

1 Tunneling

Tunneling is a phenomenon where carriers can penetrate a potential barrier due to their wave nature. In heterostructures, the band offsets at the interfaces create barriers or wells, affecting carrier tunneling across the heterojunction. The probability of tunneling is determined by the barrier height, barrier width, and the effective mass of the carriers.

The tunneling probability can be approximated using the Wentzel-Kramers-Brillouin (WKB) approximation. For a rectangular barrier, the transmission coefficient ($\mathcal{T}$) can be calculated as:

$$\mathcal{T} \approx \exp\left(-2\kappa L\right) \tag{24.5}$$

where κ is the decay constant given by:

$$\kappa = \frac{\sqrt{2m^*}}{\hbar}\sqrt{E-V} \tag{24.6}$$

Here, m^* represents the effective mass of the carrier, $\hbar$ is the reduced Planck's constant, E is the energy of the carrier, and V is the barrier height. The tunneling probability determines the conductivity and transport characteristics in heterostructures, making it a crucial factor in device performance.

2 Carrier Confinement

The energy band alignment in heterostructures results in carrier confinement, where carriers are spatially localized within certain regions due to potential barriers or wells. This confinement leads to quantum effects, influencing the carrier density of states and carrier transport properties. The confinement can be either lateral (2D) or vertical (1D), depending on the heterostructure design and the dimensions of the materials.

Lateral carrier confinement is typically achieved in quantum wells, where carriers are confined in one direction while free to move in the other two dimensions. The energy levels become quantized, resulting in discrete energy states known as subbands. The subband spacing is determined by the well width and the effective mass of the carriers.

Vertical carrier confinement is observed in heterostructures such as quantum wires or quantum dots, where carriers are confined in all three dimensions. The carrier energy levels become quantized in both the lateral and vertical directions, giving rise to discrete energy states. The carrier density of states, energy spacing, and carrier transport properties are strongly influenced by the degree of vertical confinement.

Conclusion

In this chapter, we have explored the concept of energy band alignment in heterostructures and its significance in determining transport properties in different materials. The energy band alignment at the interfaces of heterostructures plays a crucial role in carrier transport and device performance. Band offsets are used to describe the energy differences between materials and can result in either type I or type II band alignment. These alignments influence carrier confinement, tunneling, and recombination processes in heterostructures. Furthermore, the tailored band structures in heterostructures give rise to advantageous transport properties, such as carrier tunneling and confinement, enabling the development of advanced semiconductor devices with enhanced performance.

Python Code Snippet

Below is a Python code snippet that calculates energy band offsets, tunneling probability, and carrier confinement characteristics based on the concepts discussed in the chapter.

```
def calculate_band_offsets(E_v1, E_v2, E_c1, E_c2):
    '''
```

```python
    Calculate the valence and conduction band offsets for type I and
    ↪   type II heterostructures.
    :param E_v1: Valence band energy of the first material.
    :param E_v2: Valence band energy of the second material.
    :param E_c1: Conduction band energy of the first material.
    :param E_c2: Conduction band energy of the second material.
    :return: Tuple containing valence band offset and conduction
    ↪   band offset.
    '''
    delta_E_v = E_v1 - E_v2   # Valence band offset
    delta_E_c = E_c1 - E_c2   # Conduction band offset
    return delta_E_v, delta_E_c

def calculate_tunneling_probability(E, V, m_star):
    '''
    Calculate the tunneling probability using the WKB approximation.
    :param E: Energy of the carrier.
    :param V: Barrier height.
    :param m_star: Effective mass of the carrier.
    :return: Transmission coefficient (tunneling probability).
    '''
    import numpy as np
    hbar = 1.0545718e-34   # Reduced Planck's constant in J*s
    kappa = np.sqrt(2 * m_star) / hbar * np.sqrt(E - V)
    tunneling_probability = np.exp(-2 * kappa * (hbar / (2 * m_star
    ↪   * (E - V))**0.5))
    return tunneling_probability

def calculate_quantum_well_energy_levels(n, width, m_star):
    '''
    Calculate quantized energy levels in a quantum well.
    :param n: Quantum number (1, 2, 3...).
    :param width: Width of the quantum well in meters.
    :param m_star: Effective mass of the carrier in kg.
    :return: Quantized energy level in Joules.
    '''
    hbar = 1.0545718e-34   # Reduced Planck's constant in J*s
    energy_level = (n**2 * np.pi**2 * hbar**2) / (2 * m_star *
    ↪   width**2)
    return energy_level

# Example inputs for the calculations
E_v1 = 1.0   # eV
E_v2 = 0.9   # eV
E_c1 = 1.5   # eV
E_c2 = 1.2   # eV
carrier_energy = 0.95   # eV
barrier_height = 0.85   # eV
effective_mass = 9.1e-31   # kg (mass of electron)
quantum_number = 1   # First energy level
well_width = 1e-9   # 1 nm

# Calculations
```

```
delta_E_v, delta_E_c = calculate_band_offsets(E_v1, E_v2, E_c1,
 ↪ E_c2)
tunneling_probability =
 ↪ calculate_tunneling_probability(carrier_energy, barrier_height,
 ↪ effective_mass)
quantized_energy_level =
 ↪ calculate_quantum_well_energy_levels(quantum_number, well_width,
 ↪ effective_mass)

# Output results
print("Valence Band Offset (eV):", delta_E_v)
print("Conduction Band Offset (eV):", delta_E_c)
print("Tunneling Probability:", tunneling_probability)
print("Quantized Energy Level (Joules):", quantized_energy_level)
```

This code defines three functions:

- `calculate_band_offsets` computes the valence and conduction band offsets between two materials.
- `calculate_tunneling_probability` estimates the tunneling probability based on the carrier energy, barrier height, and effective mass.
- `calculate_quantum_well_energy_levels` calculates the quantized energy levels within a quantum well based on the quantum number and material properties.

The provided example calculates the band offsets, tunneling probability, and energy levels, then prints the results.

Chapter 25

Quantum Confinement

In this chapter, we explore the effects of quantum confinement in low-dimensional semiconductor structures, particularly in quantum wells. Quantum confinement refers to the phenomenon where the motion of charge carriers, such as electrons and holes, is restricted in one or more dimensions, resulting in discrete energy levels and altered transport properties. This effect plays a crucial role in the design and operation of various semiconductor devices, such as quantum well lasers and photodetectors.

Energy Quantization in Quantum Wells

Quantum wells are thin layers of semiconductor material sandwiched between two wider-bandgap materials. The quantum confinement in these structures arises due to the spatial restriction of charge carriers along the growth direction, typically denoted as the z-axis. The reduced dimensionality modifies the electronic energy levels, leading to discrete energy states known as subbands.

The energy quantization in quantum wells can be understood using the Schrödinger equation. In the simplest case of an electron confined to a one-dimensional quantum well with infinite potential barriers, the time-independent Schrödinger equation can be written as:

$$\frac{d^2\psi}{dz^2} + \frac{2m^*}{\hbar^2}(E - V(z))\psi = 0 \qquad (25.1)$$

Here, ψ represents the wave function describing the electron, E

is the energy of the electron, $V(z)$ is the potential energy, and m^* is the effective mass of the electron in the quantum well.

Solving this equation subject to appropriate boundary conditions yields the quantized energy levels, given by:

$$E_n = \frac{n^2 \pi^2 \hbar^2}{2m^* L^2} \qquad (25.2)$$

where n is the quantum number representing the energy level, $\hbar$ is the reduced Planck's constant, and L is the width of the quantum well along the confinement direction.

The energy quantization leads to the formation of discrete subbands in the energy band diagram of the quantum well. The spacing between these subbands is inversely proportional to the well width, making it possible to engineer the energy levels by controlling the dimensions of the well structure.

Density of States in Quantum Wells

The density of states (DOS) is a fundamental quantity in semiconductor physics that describes the number of available energy states in a given energy range. In quantum wells, the DOS is modified due to the energy quantization and the reduced dimensionality.

To calculate the DOS in a quantum well, we can start by considering the one-dimensional density of states $D(E)$ for a single electron in a quantum well. This is given by:

$$D(E) = \frac{m^* L}{\pi^2 \hbar^2} \qquad (25.3)$$

where m^* is the effective mass of the electron and L is the width of the quantum well. This equation represents the number of states per unit energy per unit length.

The total density of states $D_T(E)$ in a quantum well can be obtained by summing the contributions from all available electronic subbands. For a well with N_s subbands, the expression becomes:

$$D_T(E) = \sum_{n=1}^{N_s} D(E - E_n) \qquad (25.4)$$

where E_n represents the quantized energy levels derived earlier.

The modified DOS in quantum wells influences various carrier transport properties, such as carrier concentration, carrier mobility, and the optical response of the structures.

Carrier Confinement and Wavefunctions

Quantum confinement in low-dimensional semiconductor structures results in the spatial localization of charge carriers. The confinement restricts the motion of carriers along the confinement direction, leading to modified wavefunctions and carrier confinement characteristics.

In the case of electrons in a quantum well, the wavefunction $\Psi(z)$ represents the probability amplitude of finding the electron at position z along the confinement direction. The wavefunction can be calculated by solving the Schrödinger equation mentioned earlier. Each subband in the quantum well has a corresponding wavefunction associated with it.

The wavefunctions exhibit a characteristic spatial oscillation pattern within the quantum well, which depends on the quantum number n of the energy level. For higher energy levels ($n > 1$), the wavefunction exhibits more oscillations, indicating a higher spatial localization.

The carrier confinement leads to important consequences, such as the enhanced effective mass of the carriers in the quantum well and the modification of carrier scattering processes. These effects significantly impact the electrical and optical properties of the quantum well structures.

Conclusion

In this chapter, we have explored the effects of quantum confinement in low-dimensional semiconductor structures, with a specific focus on quantum wells. The confinement of charge carriers in these structures results in energy quantization, leading to discrete energy levels known as subbands. Furthermore, the density of states is modified due to the quantum confinement, which influences carrier transport properties. The spatial confinement of carriers also alters their wavefunctions, affecting their effective mass and scattering processes. The understanding of these quantum effects is essential for the design and optimization of semiconductor devices based on quantum wells.

Python Code Snippet

Below is a Python code snippet that calculates the quantized energy levels, density of states in quantum wells, and the wavefunctions of carriers under quantum confinement.

```python
import numpy as np
import matplotlib.pyplot as plt

def calculate_quantized_energy_levels(n, L, m_star):
    '''
    Calculate the quantized energy levels in a quantum well.
    :param n: Quantum number (1, 2, 3, ...).
    :param L: Width of the quantum well in meters.
    :param m_star: Effective mass of the electron in kg.
    :return: Quantized energy level in joules.
    '''
    h_bar = 1.0545718e-34  # Reduced Planck's constant in J.s
    E_n = (n**2 * np.pi**2 * h_bar**2) / (2 * m_star * L**2)
    return E_n

def density_of_states(n, m_star, L):
    '''
    Calculate the density of states for a single subband in a
    ↪ quantum well.
    :param n: Quantum number.
    :param m_star: Effective mass of the electron in kg.
    :param L: Width of the quantum well in meters.
    :return: Density of states in states per unit energy per unit
    ↪ length.
    '''
    D_E = m_star * L / (np.pi**2 * (1.0545718e-34)**2)
    return D_E

def plot_wavefunction(n, L):
    '''
    Plot the wavefunction for a given quantum number in a quantum
    ↪ well.
    :param n: Quantum number.
    :param L: Width of the quantum well in meters.
    '''
    z = np.linspace(0, L, 1000)
    # Use the wavefunction for the infinite square well
    psi = np.sqrt(2 / L) * np.sin(n * np.pi * z / L)

    plt.figure(figsize=(10, 5))
    plt.plot(z, psi**2, label=f'|(z)|² for n={n}')
    plt.title('Wavefunction Probability Density in Quantum Well')
    plt.xlabel('Position (z) [m]')
    plt.ylabel('Probability Density |(z)|²')
    plt.grid()
```

```
    plt.legend()
    plt.show()

# Inputs for the calculations
n_values = [1, 2, 3]   # First three quantum levels
L = 1e-9   # Width of the quantum well in meters (1 nm)
m_star = 9.11e-31   # Effective mass of the electron in kg

# Calculating quantized energy levels
for n in n_values:
    energy = calculate_quantized_energy_levels(n, L, m_star)
    print(f"Quantized Energy Level (n={n}): {energy:.3e} J")

# Density of states for the first subband
D_E = density_of_states(1, m_star, L)
print(f"Density of States for first subband: {D_E:.3e} states/eV.m")

# Plot wavefunction for the first three energy levels
for n in n_values:
    plot_wavefunction(n, L)
```

This code defines three functions:

- `calculate_quantized_energy_levels` computes the quantized energy levels for electrons in a quantum well using the equation $E\_n = \frac{n^2 \pi^2 \hbar^2}{2m^* L^2}$.
- `density_of_states` calculates the one-dimensional density of states using the formula $D(E) = \frac{m^* L}{\pi^2 \hbar^2}$.
- `plot_wavefunction` generates and displays the probability density of the wavefunction for the specified quantum numbers in a quantum well.

The provided example computes and prints the quantized energy levels for the first three subbands and the density of states for the first subband, followed by visualizing the wavefunctions for these energy levels.

Chapter 26

Thermal Effects

In this chapter, we delve into the intricate relationship between temperature and semiconductor behavior, with a particular emphasis on thermal conductivity. Temperature plays a crucial role in determining the performance and reliability of semiconductor devices. The understanding of thermal effects is essential for designing and optimizing semiconductor-based systems.

Thermal Conductivity and Heat Transfer

Thermal conductivity is a fundamental property that characterizes the ability of a material to conduct heat. It quantifies the rate of heat transfer per unit area per unit temperature gradient. In the context of semiconductors, thermal conductivity is a key parameter that governs the dissipation of heat generated during device operation.

The thermal conductivity of a semiconductor material, often denoted as κ, depends on several factors, including its atomic composition, crystal structure, and temperature. At low temperatures, where phonon scattering dominates, the thermal conductivity is primarily determined by phonon-phonon interactions. On the other hand, at elevated temperatures, additional scattering mechanisms, such as phonon-electron and phonon-impurity interactions, start to play a significant role.

Mathematically, the heat conduction equation for a semiconductor material can be expressed as:

$$\frac{dQ}{dt} = -\kappa A \frac{dT}{dx} \qquad (26.1)$$

where $\frac{dQ}{dt}$ represents the rate of heat transfer in watts, κ is the thermal conductivity in watts per meter-kelvin (W/m-K), A is the cross-sectional area perpendicular to the heat flow, and $\frac{dT}{dx}$ is the temperature gradient in kelvin per meter (K/m).

Temperature Dependence of Thermal Conductivity

The thermal conductivity of semiconductors exhibits a strong temperature dependence. As temperature increases, various scattering mechanisms become more pronounced, leading to a reduction in thermal conductivity.

In a crystalline semiconductor, phonons are the primary heat carriers. The relationship between thermal conductivity and temperature can be described by the Callaway model, which takes into account the scattering of phonons by lattice defects, impurities, and phonon-phonon interactions. The Callaway model predicts that the thermal conductivity κ of a semiconductor material decreases with increasing temperature T according to the relationship:

$$\kappa = \kappa_0 - \frac{BT^m}{\Theta_D} \qquad (26.2)$$

where κ_0 is the residual thermal conductivity at absolute zero temperature, B is a temperature-dependent constant, m is the temperature exponent, and Θ_D is the characteristic Debye temperature.

The temperature exponent m in the Callaway model varies between 1 and 2, depending on the dominant scattering mechanisms. Typically, for crystalline semiconductors, m is close to 1 at low temperatures, indicating phonon-phonon scattering dominance, while it approaches 2 at high temperatures, where phonon-electron and phonon-impurity scattering become prevalent.

Impact on Semiconductor Behavior

Temperature has a profound impact on the electrical and optical properties of semiconductor materials. As temperature increases, several phenomena manifest, leading to alterations in the semiconductor behavior.

Firstly, as temperature rises, the carrier concentration in semiconductors generally increases due to enhanced thermal excitation of electrons from the valence band to the conduction band. This phenomenon is quantitatively described by the intrinsic carrier concentration n_i, which follows an exponential relationship with temperature, given by:

$$n_i = AT^{3/2} e^{-E_g/(2k_B T)} \tag{26.3}$$

where A is a temperature-independent constant, E_g is the bandgap energy of the semiconductor material, and k_B is the Boltzmann constant.

Furthermore, the mobility of charge carriers in semiconductors is often inversely related to temperature. The reduction in mobility arises from increased scattering due to lattice vibrations and impurity interactions, leading to a decrease in the average carrier drift velocity.

The variation of semiconductor properties with temperature affects the performance and reliability of semiconductor devices, such as transistors and diodes. Understanding and modeling these thermal effects are critical for proper device operation and design optimization.

Thermal Conductivity Measurements

Experimental determination of the thermal conductivity is vital for material characterization and device development. Various techniques have been developed to measure the thermal conductivity of semiconductor materials as a function of temperature.

One commonly used method is the steady-state heat flow method, which involves measuring the temperature difference across a sample with a known cross-sectional area and length subjected to a constant heat source. The thermal conductivity can be calculated using Fourier's law of heat conduction and the known heat power applied.

Another technique is the transient thermoreflectance method, which is based on laser-induced heating of the sample and subsequent measurement of the temperature response using optical reflectivity. This method allows for high time resolution measurements and is suitable for thin film and nanoscale samples.

Other advanced experimental techniques, such as the 3 method and the photothermal deflection method, have also been employed to measure the thermal conductivity of semiconductors with high accuracy and sensitivity.

Conclusion

This chapter has provided a comprehensive overview of the impact of temperature on semiconductor behavior and thermal conductivity. We have explored the temperature dependence of thermal conductivity in semiconductors, considering phonon-phonon, phonon-electron, and phonon-impurity scattering mechanisms. The influence of temperature on carrier concentration and mobility in semiconductors has been highlighted. Furthermore, experimental techniques for measuring thermal conductivity have been discussed. The understanding of thermal effects is critical for characterizing semiconductor materials accurately and optimizing the performance of semiconductor devices.

Python Code Snippet

Below is a Python code snippet that calculates thermal conductivity, intrinsic carrier concentration, and models temperature dependence based on the important equations and algorithms mentioned in this chapter.

```python
def calculate_thermal_conductivity(kappa_0, B, T, Theta_D, m):
    '''
    Calculate the thermal conductivity of a semiconductor based on
      temperature.
    :param kappa_0: Residual thermal conductivity at absolute zero
      temperature (W/m-K).
    :param B: Temperature-dependent constant.
    :param T: Temperature in Kelvin.
    :param Theta_D: Characteristic Debye temperature (K).
    :param m: Temperature exponent (1 or 2).
    :return: Thermal conductivity (W/m-K).
    '''
    kappa = kappa_0 - (B * (T ** m)) / Theta_D
    return kappa

def calculate_intrinsic_carrier_concentration(A, E_g, T):
    '''
```

```python
    Calculate the intrinsic carrier concentration of a
    ↪ semiconductor.
    :param A: Temperature-independent constant.
    :param E_g: Bandgap energy of the semiconductor material (eV).
    :param T: Temperature in Kelvin.
    :return: Intrinsic carrier concentration (m^-3).
    '''
    import math
    k_B = 8.617333e-5   # Boltzmann constant in eV/K
    n_i = A * (T ** (3/2)) * math.exp(-E_g / (2 * k_B * T))
    return n_i


# Inputs for the calculations
kappa_0 = 200   # Residual thermal conductivity (W/m-K)
B = 1.0e-10   # Temperature-dependent constant
Theta_D = 300   # Characteristic Debye temperature (K)
E_g = 1.12   # Bandgap energy of silicon (eV)
A = 1.08e6   # Temperature-independent constant
temperatures = [100, 200, 300, 400]   # Temperatures in Kelvin

# Calculate and print thermal conductivity at different temperatures
print("Thermal Conductivity Calculations:")
for T in temperatures:
    thermal_conductivity = calculate_thermal_conductivity(kappa_0,
    ↪ B, T, Theta_D, m=1)
    print(f"At {T} K, Thermal Conductivity:
    ↪ {thermal_conductivity:.2f} W/m-K")

# Calculate and print intrinsic carrier concentration at different
↪ temperatures
print("\nIntrinsic Carrier Concentration Calculations:")
for T in temperatures:
    intrinsic_concentration =
    ↪ calculate_intrinsic_carrier_concentration(A, E_g, T)
    print(f"At {T} K, Intrinsic Carrier Concentration:
    ↪ {intrinsic_concentration:.2e} m^-3")
```

This code defines two functions:

- `calculate_thermal_conductivity` computes the thermal conductivity of a semiconductor based on the residual conductivity, constants, temperature, and the characteristic Debye temperature.
- `calculate_intrinsic_carrier_concentration` determines the intrinsic carrier concentration considering the bandgap energy, a temperature-independent constant, and temperature.

The provided example calculates the thermal conductivity and the intrinsic carrier concentration at specified temperatures, printing the results for each calculation.

Chapter 27

Noise Models

In this chapter, we explore the sources and mathematical models for noise in semiconductor devices. Noise refers to the unwanted fluctuations or disturbances that can degrade the performance and reliability of electronic systems. Understanding and accurately modeling noise in semiconductors is essential for designing and optimizing devices for various applications.

Sources of Noise in Semiconductor Devices

Semiconductor devices are susceptible to different sources of noise, including thermal noise, shot noise, flicker or 1/f noise, and quantum noise. Each type of noise arises from different physical processes and has distinct characteristics.

1 Thermal Noise

Thermal noise, also known as Johnson-Nyquist noise, originates from the random motion of charge carriers due to thermal energy. It is present even at absolute zero temperature and follows a Gaussian distribution. The power spectral density (PSD) of thermal noise, in units of volts squared per hertz (V^2/Hz), is given by the equation:

$$S_{\text{thermal}}(f) = 4k_B TR$$

where k_B is the Boltzmann constant, T is the temperature in Kelvin, and R is the resistance of the device.

2 Shot Noise

Shot noise arises due to the discrete nature of electric current, caused by random fluctuations in the arrival rate of individual charge carriers. It is particularly significant in devices with low currents or at high frequencies. The PSD of shot noise, in units of ampere squared per hertz (A2/Hz), is given by:

$$S_{\text{shot}}(f) = 2qI$$

where q is the charge of an electron and I is the current.

3 Flicker or 1/f Noise

Flicker noise, also known as 1/f noise, has a spectral density that decreases with increasing frequency. It is observed in many electronic components and is often attributed to variations in carrier mobility or trap-assisted processes in the device. The PSD of flicker noise, in units of volts squared per hertz (V^2/Hz), can be modeled as:

$$S_{\text{flicker}}(f) = K \left(\frac{1}{f}\right)^\alpha$$

where K is a constant and α is the noise exponent, typically between 0.5 and 2.

4 Quantum Noise

Quantum noise arises due to the inherent probabilistic nature of quantum mechanical processes. It includes shot noise, as well as other noise sources related to the quantum behavior of charge carriers. Quantum noise is especially relevant in devices operating at low temperatures or with small dimensions.

Noise Models

Mathematical models are used to characterize and predict the behavior of noise in semiconductor devices. These models provide insights into noise characteristics and enable the assessment of noise performance.

1 Power Spectral Density (PSD)

The power spectral density (PSD) is a fundamental tool for analyzing noise in semiconductor devices. It represents the power content of noise as a function of frequency. The PSD of a noise signal is typically measured experimentally or estimated using mathematical techniques such as Fourier analysis.

2 Nyquist Formula

The Nyquist formula relates the PSD of thermal noise to the temperature and resistance of a device. It provides a quantitative expression for the expected noise level in a resistor at a given temperature. According to the Nyquist formula, the PSD of thermal noise can be calculated using:

$$S_{\text{thermal}}(f) = 4k_B T R$$

where k_B is the Boltzmann constant, T is the temperature in Kelvin, and R is the resistance.

3 White Noise Model

The white noise model is a simplifying assumption that represents noise as a stationary random process with constant power spectral density (PSD) across all frequencies. While this model does not capture the frequency-dependent characteristics of most types of noise, it serves as a useful approximation in many applications.

4 Correlation Function

The correlation function provides a mathematical tool to analyze the statistical properties of noise signals. It measures the similarity between two points in a random process as a function of time. The correlation function is often used to determine the correlation time, which quantifies the time scale at which noise signals are no longer correlated.

5 Noise Figure

The noise figure is a metric used to quantify the degradation in the signal-to-noise ratio caused by a device or circuit. It measures the additional noise introduced by the device compared to an ideal noiseless device. The noise figure is usually expressed in decibels (dB) and represents the ratio of the output noise power of the device to the noise power that would be generated by an ideal noiseless device.

Conclusion

In this chapter, we have examined the sources of noise in semiconductor devices, including thermal noise, shot noise, flicker or 1/f noise, and quantum noise. The mathematical models and techniques used to characterize and model noise have also been discussed, such as the power spectral density (PSD), Nyquist formula, white noise model, correlation function, and noise figure. These models and methodologies are essential for understanding, analyzing, and designing semiconductor devices that operate reliably in the presence of noise.

Python Code Snippet

Below is a Python code snippet that implements the key equations and models related to noise in semiconductor devices as discussed in this chapter.

```python
import numpy as np
import matplotlib.pyplot as plt

def thermal_noise_psd(T, R):
    '''
    Calculate the power spectral density of thermal noise
    ↪ (Johnson-Nyquist noise).
    :param T: Temperature in Kelvin.
    :param R: Resistance in ohms.
    :return: Power spectral density in V^2/Hz.
    '''
    k_B = 1.380649e-23  # Boltzmann constant in J/K
    return 4 * k_B * T * R

def shot_noise_psd(I):
    '''
    Calculate the power spectral density of shot noise.
    :param I: Current in amperes.
    :return: Power spectral density in A^2/Hz.
    '''
    q = 1.602176634e-19  # Charge of an electron in coulombs
    return 2 * q * I
```

```python
def flicker_noise_psd(f, K, alpha):
    '''
    Calculate the power spectral density of flicker noise (1/f
    ↪ noise).
    :param f: Frequency in hertz.
    :param K: Constant specific to the device.
    :param alpha: Noise exponent.
    :return: Power spectral density in V^2/Hz.
    '''
    return K * (1 / f) ** alpha

def generate_noise_signals(T, R, I, K, alpha, frequency_range):
    '''
    Generate and plot the power spectral density for different noise
    ↪ sources.
    :param T: Temperature in Kelvin.
    :param R: Resistance in ohms.
    :param I: Current in amperes.
    :param K: Constant specific to the device for flicker noise.
    :param alpha: Noise exponent for flicker noise.
    :param frequency_range: List of frequencies in hertz.
    '''
    psd_thermal = [thermal_noise_psd(T, R) for f in frequency_range]
    psd_shot = [shot_noise_psd(I) for f in frequency_range]
    psd_flicker = [flicker_noise_psd(f, K, alpha) for f in
    ↪ frequency_range]

    plt.figure(figsize=(10, 6))
    plt.semilogx(frequency_range, psd_thermal, label='Thermal Noise
    ↪ PSD', color='blue')
    plt.semilogx(frequency_range, psd_shot, label='Shot Noise PSD',
    ↪ color='green')
    plt.semilogx(frequency_range, psd_flicker, label='Flicker Noise
    ↪ PSD', color='orange')
    plt.title('Power Spectral Density of Noise Sources')
    plt.xlabel('Frequency (Hz)')
    plt.ylabel('Power Spectral Density (V^2/Hz or A^2/Hz)')
    plt.legend()
    plt.grid()
    plt.show()

# Inputs for the calculations
T = 300    # Temperature in Kelvin (room temperature)
R = 1000   # Resistance in ohms
I = 1e-6   # Current in amperes (1 microampere)
K = 1e-12  # Constant for flicker noise
alpha = 1  # Noise exponent (typical value)
frequency_range = np.logspace(1, 6, num=100)  # Frequency range from
↪ 10 Hz to 1 MHz

# Generate and plot noise signals
```

`generate_noise_signals(T, R, I, K, alpha, frequency_range)`

This code defines several functions:

- `thermal_noise_psd` calculates the power spectral density of thermal noise based on temperature and resistance.
- `shot_noise_psd` computes the power spectral density of shot noise based on the current.
- `flicker_noise_psd` models the power spectral density of flicker noise using frequency, a constant, and a noise exponent.
- `generate_noise_signals` combines these functions to generate and plot the power spectral density for thermal, shot, and flicker noise across a specified frequency range.

The provided example calculates and visualizes the power spectral density of different noise sources affecting semiconductor devices, enabling a better understanding of their impact on performance.

Chapter 28

Carrier Recombination-Generation

This chapter delves into the intricate mechanisms underlying carrier recombination and generation in semiconductors, exploring their profound effects on semiconductor performance. By understanding the processes through a mathematical lens, we can gain insights into the behavior of carriers and optimize the design and operation of semiconductor devices.

1 Introduction

Carrier recombination and generation are fundamental processes that significantly impact the behavior of semiconductors. Recombination refers to the process in which charge carriers combine, leading to a decrease in their population. Conversely, generation involves the creation of charge carriers, resulting in an increase in their concentration. Both processes have a profound influence on the carrier density and lifetime, thus affecting the overall performance of semiconductor devices.

2 Rate Equations

Mathematically describing carrier recombination and generation requires the formulation of rate equations that govern the dynamics of carriers' concentrations. These equations quantify the rate at which carriers recombine or generate and provide insights into the underlying mechanisms.

The rate equation for recombination is given by

$$\frac{dn}{dt} = -\frac{n}{\tau_n},$$

where n represents the concentration of charge carriers and τ_n denotes the carrier lifetime.

Likewise, the rate equation for generation is expressed as

$$\frac{dn}{dt} = \frac{G}{\tau_p} - \frac{n}{\tau_p},$$

where G represents the rate of carrier generation and τ_p denotes the carrier lifetime.

3 Detailed Balance Equation

The detailed balance equation provides a fundamental principle for understanding the relationship between carrier recombination and generation. It states that in thermal equilibrium, the rate of carrier recombination is equal to the rate of carrier generation. Mathematically, this can be expressed as

$$R = G,$$

where R represents the recombination rate and G represents the generation rate.

4 Recombination Mechanisms

Various recombination mechanisms contribute to carrier recombination in semiconductors. Three primary recombination processes are widely studied:

1. *Shockley-Read-Hall (SRH) Recombination*: This mechanism involves the interaction of charge carriers with trapping centers, which capture and subsequently release them. The SRH recombination rate equation is given by

$$R_{\text{SRH}} = \frac{np - n_i^2}{\tau_{\text{SRH}}},$$

where n and p represent the concentrations of electrons and holes, respectively, and τ_{SRH} denotes the SRH recombination lifetime.

2. *Radiative Recombination*: In this process, charge carriers recombine through the emission of photons. The radiative recombination rate equation can be expressed as

$$R_{\text{radiative}} = B_{\text{rad}} np,$$

where B_{rad} represents the radiative recombination coefficient.

3. *Auger Recombination*: Auger recombination occurs when one charge carrier transfers its energy to another carrier, resulting in the recombination of both carriers. The Auger recombination rate equation is given by

$$R_{\text{Auger}} = C_{\text{Auger}} n^2 p,$$

where C_{Auger} denotes the Auger recombination coefficient.

5 Generation Mechanisms

Carrier generation in semiconductors arises from various processes, including:

1. *Thermal Generation*: Thermal energy can excite electrons from the valence band to the conduction band, resulting in carrier generation. The thermal generation rate equation is given by

$$G_{\text{thermal}} = A_{\text{thermal}}(np - n_i^2) e^{-\frac{E_g}{kT}},$$

where A_{thermal} represents the thermal generation coefficient, E_g denotes the band gap energy, k is the Boltzmann constant, and T denotes the temperature.

2. *Recombination-Enhanced Generation*: In certain situations, recombination processes can enhance carrier generation. The resulting generation rate equation can be expressed as

$$G_{\text{recomb-enhanced}} = C_{\text{rec}} R,$$

where C_{rec} represents the recombination-enhanced generation coefficient and R denotes the recombination rate.

3. *Photon Generation*: Photons absorbed within a semiconductor can generate electron-hole pairs through the photoelectric effect. The photon generation rate equation is given by

$$G_{\text{photon}} = B_{\text{photon}} I_{\text{light}},$$

where B_{photon} represents the photon generation coefficient and I_{light} denotes the incident light intensity.

6 Impacts on Semiconductor Performance

Carrier recombination and generation have significant implications for the overall performance of semiconductors. They influence key parameters such as carrier density, lifetime, and transport efficiency, ultimately affecting the device's operation and characteristics.

Excess carrier recombination leads to a decrease in carrier lifetime and an overall reduction in device efficiency. On the other hand, efficient carrier generation is crucial for ensuring optimal device performance. By understanding and controlling these processes, we can enhance device efficiency, improve response times, and minimize losses in various semiconductor applications.

In this chapter, we explored carrier recombination and generation mechanisms in semiconductors. The rate equations provided a mathematical foundation for understanding the dynamics of carrier concentrations. Furthermore, the detailed balance equation highlighted the equilibrium between recombination and generation rates. Various recombination and generation mechanisms, such as SRH, radiative, Auger, thermal, recombination-enhanced, and photon generation, were described, shedding light on their respective impact on semiconductor performance.

Python Code Snippet

Below is a Python code snippet that implements important equations and algorithms mentioned in the chapter regarding carrier recombination and generation in semiconductors.

```python
def carrier_recombination(n, p, n_i, tau_SRH):
    '''
    Calculate the Shockley-Read-Hall recombination rate.
    :param n: Concentration of electrons (cm^-3).
    :param p: Concentration of holes (cm^-3).
    :param n_i: Intrinsic carrier concentration (cm^-3).
    :param tau_SRH: SRH recombination lifetime (seconds).
    :return: Recombination rate (cm^-3 s^-1).
    '''
    return (n * p - n_i**2) / tau_SRH

def radiative_recombination(n, p, B_rad):
    '''
```

```python
    Calculate the radiative recombination rate.
    :param n: Concentration of electrons (cm^-3).
    :param p: Concentration of holes (cm^-3).
    :param B_rad: Radiative recombination coefficient (cm^3 s^-1).
    :return: Recombination rate (cm^-3 s^-1).
    '''
    return B_rad * n * p


def auger_recombination(n, p, C_Auger):
    '''
    Calculate the Auger recombination rate.
    :param n: Concentration of electrons (cm^-3).
    :param p: Concentration of holes (cm^-3).
    :param C_Auger: Auger recombination coefficient (cm^6 s^-1).
    :return: Recombination rate (cm^-3 s^-1).
    '''
    return C_Auger * n**2 * p


def thermal_generation(A_thermal, n_i, E_g, k, T):
    '''
    Calculate the thermal generation rate.
    :param A_thermal: Thermal generation coefficient.
    :param n_i: Intrinsic carrier concentration (cm^-3).
    :param E_g: Band gap energy (eV).
    :param k: Boltzmann constant (eV/K).
    :param T: Temperature (K).
    :return: Generation rate (cm^-3 s^-1).
    '''
    return A_thermal * (n_i**2) * np.exp(-E_g / (k * T))


def photon_generation(B_photon, I_light):
    '''
    Calculate the photon generation rate.
    :param B_photon: Photon generation coefficient (cm^-2 s^-1).
    :param I_light: Incident light intensity (W/cm^2).
    :return: Generation rate (cm^-3 s^-1).
    '''
    return B_photon * I_light


# Example Constants
n = 1e10      # Electron concentration in cm^-3
p = 1e10      # Hole concentration in cm^-3
n_i = 1.5e10  # Intrinsic carrier concentration in cm^-3
tau_SRH = 1e-6   # SRH lifetime in seconds
B_rad = 1e-10    # Radiative recombination coefficient in cm^3 s^-1
C_Auger = 1e-30  # Auger recombination coefficient in cm^6 s^-1
A_thermal = 1e-14   # Thermal generation coefficient
E_g = 1.12    # Band gap energy in eV (Silicon)
k = 8.617e-5  # Boltzmann constant in eV/K
```

```
T = 300        # Temperature in K
B_photon = 1e-4    # Photon generation coefficient in cm^-2 s^-1
I_light = 0.1    # Incident light intensity in W/cm^2

# Calculations
R_SRH = carrier_recombination(n, p, n_i, tau_SRH)
R_rad = radiative_recombination(n, p, B_rad)
R_Auger = auger_recombination(n, p, C_Auger)
G_thermal = thermal_generation(A_thermal, n_i, E_g, k, T)
G_photon = photon_generation(B_photon, I_light)

# Output results
print("Shockley-Read-Hall Recombination Rate:", R_SRH, "cm^-3 s^-1")
print("Radiative Recombination Rate:", R_rad, "cm^-3 s^-1")
print("Auger Recombination Rate:", R_Auger, "cm^-3 s^-1")
print("Thermal Generation Rate:", G_thermal, "cm^-3 s^-1")
print("Photon Generation Rate:", G_photon, "cm^-3 s^-1")
```

This code defines five functions:

- carrier_recombination calculates the Shockley-Read-Hall recombination rate based on electron and hole concentrations.
- radiative_recombination computes the radiative recombination rate given the concentrations and coefficient.
- auger_recombination calculates the Auger recombination rate based on carrier densities and coefficient.
- thermal_generation computes the thermal generation rate as a function of temperature and band gap energy.
- photon_generation calculates the rate of carrier generation due to photon absorption.

The provided code serves to compute various recombination and generation rates, which are essential for understanding semiconductor performance, and then prints the results.

Chapter 29

Non-Equilibrium Green's Function (NEGF)

In this chapter, we delve into the mathematical framework and approach of Non-Equilibrium Green's Function (NEGF) for modeling quantum transport in semiconductor devices. The NEGF method provides a powerful tool for analyzing the behavior of electrons and holes under non-equilibrium conditions, allowing us to gain a deeper understanding of the dynamics and performance of these devices. We explore the key concepts and equations involved in NEGF, emphasizing their significance in the field of semiconductor physics.

1 Introduction to NEGF

The Non-Equilibrium Green's Function (NEGF) method is a quantum mechanical approach widely utilized for analyzing the transport properties of semiconductor devices. This method enables us to study the behavior of electrons and holes in these devices and understand their flow and interaction under non-equilibrium conditions.

2 Fundamentals of NEGF

The NEGF approach is based on the formalism of Green's functions, which are mathematical objects that provide a description of the quantum mechanical behavior of particles. In the context of semiconductor physics, the NEGF formalism focuses on the Green's functions that govern electron and hole propagation and interaction.

3 Dyson's Equation

A central equation in the NEGF method is Dyson's equation, which relates the Green's function of the system to the Hamiltonian and self-energy. Mathematically, Dyson's equation can be expressed as:

$$G = G_0 + G_0 \Sigma G$$

where G is the full Green's function of the system, G_0 is the Green's function of the isolated subsystem, and Σ represents the self-energy.

4 Electron and Hole Propagation

The NEGF approach allows us to calculate the propagation of electrons and holes in semiconductor devices. By solving Dyson's equation, we can obtain the Green's function and analyze the transport of these carriers under non-equilibrium conditions.

5 Self-Energy

The self-energy term in Dyson's equation represents the interactions of electrons and holes with the surrounding system. It accounts for various scattering processes and reflects the effects of the device structure and external potentials on carrier transport.

6 Transport Equations

NEGF provides a set of equations, known as transport equations, that describe the behavior of electrons and holes in a device. These equations link the Green's functions and self-energy to the related physical quantities, such as current, voltage, and density of states.

7 Landauer-Büttiker Formalism

The Landauer-Büttiker formalism is a powerful tool within the NEGF method for calculating transport properties, such as conductance and current, in semiconductor devices. This formalism relates the Green's functions and self-energy to the scattering matrix, providing a direct connection between the quantum mechanical behavior of carriers and measurable quantities.

8 Applications of NEGF

NEGF has found numerous applications in the field of semiconductor physics. It is widely used to model and analyze the behavior of various devices, such as transistors, diodes, and quantum dot structures. NEGF allows for a detailed understanding of device performance, including aspects such as current-voltage characteristics, charge density, and quantum confinement effects.

9 Challenges and Future Directions

While NEGF is a powerful approach for modeling quantum transport in semiconductor devices, it does pose certain challenges. The computational complexity of solving Dyson's equation and obtaining accurate results can be demanding. Efforts are continually being made to develop more efficient algorithms and techniques for tackling these challenges, enhancing the applicability of NEGF to real-world device simulations.

In this chapter, we have explored the mathematical framework and approach of Non-Equilibrium Green's Function (NEGF) for modeling quantum transport in semiconductor devices. We have discussed the fundamental concepts of NEGF, including the use of Green's functions and Dyson's equation. Additionally, we have touched upon the self-energy term, transport equations, and the Landauer-Büttiker formalism. NEGF has proven to be a valuable tool for analyzing and understanding the behavior of electrons and holes in semiconductor devices, enabling us to optimize device performance and explore new device architectures. Ongoing research focuses on addressing computational challenges and further advancing the capabilities of NEGF in the realm of semiconductor physics.

Python Code Snippet

Below is a Python code snippet that implements the key equations and algorithms involved in the Non-Equilibrium Green's Function (NEGF) framework for modeling quantum transport in semiconductor devices.

```python
import numpy as np

def calculate_green_function(h, sigma, energy):
    '''
    Calculate the Green's function using Dyson's equation.
    :param h: Hamiltonian matrix of the system.
    :param sigma: Self-energy matrix.
    :param energy: Energy value at which to calculate the Green's
    ↪ function.
    :return: Green's function matrix at specified energy.
    '''
    identity_matrix = np.eye(h.shape[0])
    g0 = np.linalg.inv(energy * identity_matrix - h)  # Isolated
    ↪ system Green's function
    green_function = g0 + g0 @ sigma @ g0  # Dyson's equation
    return green_function

def calculate_current(green_function, fermi_energy):
    '''
    Calculate the current using the Green's function and the
    ↪ Fermi-Dirac distribution.
    :param green_function: Green's function matrix.
    :param fermi_energy: Fermi energy level of the system.
    :return: Current value.
    '''
    current = 0
    energies = np.linspace(-10, 10, 100)  # Energy range for
    ↪ calculation
    for energy in energies:
        # Calculate the Fermi-Dirac distribution
        fd_distribution = 1 / (np.exp((energy - fermi_energy) /
        ↪ (8.617e-5)) + 1)  # eV to Kelvin conversion
        current += np.trace(green_function @
        ↪ np.diag(fd_distribution))
    return current

def main():
    # Define system parameters
    h_matrix = np.array([[1, 0.5], [0.5, 2]])  # Example
    ↪ Hamiltonian
    sigma_matrix = np.array([[0.1, 0], [0, 0.1]])  # Example
    ↪ self-energy
    energy_value = 1.0  # Example energy value in eV
    fermi_energy_value = 0.5  # Example Fermi energy in eV
```

```python
    # Calculate the Green's function
    green_func = calculate_green_function(h_matrix, sigma_matrix,
     ↪ energy_value)

    # Calculate the current
    current_value = calculate_current(green_func,
     ↪ fermi_energy_value)

    # Output results
    print("Green's Function Matrix:\n", green_func)
    print("Current: ", current_value)

if __name__ == "__main__":
    main()
```

This code defines the following functions and their purposes:

- `calculate_green_function` calculates the Green's function using Dyson's equation based on the Hamiltonian and self-energy matrices at a specified energy.
- `calculate_current` computes the current in the system using the obtained Green's function and the Fermi-Dirac distribution.
- `main` sets up the system parameters, performs the calculations, and prints the resulting Green's function and current.

The provided example models a simple system defined by a Hamiltonian and calculates its Green's function and resultant current under non-equilibrium conditions.

Chapter 30
Phonon Interaction

In this chapter, we explore the impact of lattice vibrations, known as phonons, on the electrical properties of semiconductors. Phonons play a crucial role in determining the thermal conductivity, carrier mobility, and other quantum transport phenomena in these materials. We will delve into the mathematical modeling of phonon interactions and their effects, providing insight into the intricate relationship between lattice vibrations and semiconductor behavior.

1 Introduction to Phonons

Phonons are quantized vibrations of the crystal lattice in a solid material. They can be viewed as packets of lattice energy, propagating through the crystal structure. These vibrations are crucial in determining various physical properties of semiconductors, including their thermal and electrical conductivities.

2 Lattice Dynamics

To understand the impact of phonons on semiconductor behavior, we must delve into the mathematical study of lattice dynamics. This field focuses on modeling the vibrational modes and energies of the crystal lattice. The lattice dynamics of semiconductors can be described using techniques such as the Solid State Physics concept of normal modes.

3 Normal Modes

The normal modes of a crystal lattice refer to the collective vibrational patterns of its constituent atoms. By solving the appropriate equations of motion, we can determine the frequencies and wavevectors associated with these normal modes. The normal modes form the basis for understanding phonon interactions and their impact on semiconductor properties.

4 Phonon Dispersion

Phonon dispersion describes the relationship between the phonon frequency and wavevector in a crystal lattice. By considering the periodicity and symmetries of the lattice, we can derive the phonon dispersion relation. This relation provides insight into the allowed phonon frequencies and their dependence on the crystal structure.

5 Phonon-Phonon Interactions

An important aspect of phonons in semiconductors is their interaction with one another. Phonon-phonon interactions can cause scattering and relaxation processes, leading to changes in phonon lifetimes and energy distributions. The modeling of these interactions relies on techniques such as the Boltzmann Transport Equation and the concept of phonon scattering rates.

6 Thermal Conductivity

The interaction between phonons significantly influences the thermal conductivity of semiconductors. Phonon scattering processes, such as Umklapp scattering and boundary scattering, introduce resistance to the flow of lattice heat, reducing the overall thermal conductivity. The thermal conductivity can be quantitatively described using the phonon mean free path and the concept of thermal resistivity.

7 Phonons and Carrier Mobility

Phonons also affect the carrier mobility in semiconductors. Through scattering mechanisms, phonons can hinder the movement of charged carriers, leading to reductions in carrier mobility. The impact of phonons on carrier mobility is essential in understanding the electrical behavior of semiconductors, especially at elevated temperatures.

8 Phonon Drag

Phonon drag, also known as the Nernst-Ettingshausen effect, refers to the generation of an electric field due to the interaction between phonons and charge carriers moving across a temperature gradient. This effect is particularly significant in low-dimensional semiconductor structures, such as quantum wells and nanowires. Phonon drag can provide insights into the interplay between lattice vibrations and electrical transport.

9 Thermoelectric Effects

The interplay between phonons and charge carriers has extensive implications for thermoelectric materials and devices. The Seebeck effect, Peltier effect, and Thomson effect are examples of thermoelectric phenomena influenced by phonons. Understanding these effects is crucial in designing efficient thermoelectric devices for energy conversion applications.

10 Phonon Engineering

The impact of phonons on semiconductor properties has led to the emerging field of phonon engineering. Researchers aim to manipulate phonon behavior and design materials with tailored phonon properties to enhance thermal and electrical performance. Techniques such as nanostructuring, interface engineering, and phonon bandgap engineering are actively explored for improving semiconductor devices.

In this chapter, we have explored the impact of lattice vibrations, or phonons, on the electrical properties of semiconductors. We discussed the role of phonons in determining thermal conductivity, carrier mobility, and other transport phenomena. The mathematical study of lattice dynamics, normal modes, and phonon dispersion provides insight into the behavior of phonons in semiconductor materials. Further, we explored the effects of phonon-phonon interactions on thermal conductivity and carrier mobility. Phonon drag and thermoelectric effects were also examined, highlighting the interplay between lattice vibrations and electrical transport. Finally, the emerging field of phonon engineering was introduced, highlighting the opportunities for tailoring phonon properties in semiconductor materials.

"'latex

Python Code Snippet

Below is a Python code snippet that calculates the thermal conductivity and carrier mobility based on phonon interactions.

```python
def phonon_dispersion_relation(k, a, omega_zero):
    '''
    Calculate the phonon dispersion relation for a 1D crystal
        lattice.
    :param k: Wavevector in reciprocal space.
    :param a: Lattice constant in meters.
    :param omega_zero: Base frequency at k=0.
    :return: Frequency of phonons associated with the given
        wavevector k.
    '''
    return omega_zero * (1 - (k * a / (3.14159))**2)

def calculate_thermal_conductivity(mean_free_path, specific_heat,
    velocity):
    '''
    Calculate the thermal conductivity of a semiconductor.
    :param mean_free_path: Mean free path of phonons in meters.
    :param specific_heat: Specific heat capacity in J/(kg*K).
    :param velocity: Average velocity of phonons in m/s.
    :return: Thermal conductivity in W/(m*K).
    '''
    return (1/3) * specific_heat * velocity * mean_free_path

def calculate_carrier_mobility(phonon_scattering_time, temperature):
    '''
    Calculate the mobility of charge carriers influenced by phonons.
    :param phonon_scattering_time: Average time between phonon
        scattering events in seconds.
    :param temperature: Temperature in Kelvin.
    :return: Carrier mobility in m^2/(V*s).
    '''
    e_charge = 1.602e-19  # Charge of an electron in coulombs
    return e_charge * phonon_scattering_time / (1.38e-23 *
        temperature)

# Parameters for the calculations
lattice_constant = 1e-9  # Lattice constant in meters
omega_zero = 1e13  # Base frequency in Hz
```

```
wavevector = 1e6  # Wavevector in m^-1
mean_free_path = 1e-6  # Mean free path in meters
specific_heat = 1000  # Specific heat in J/(kg*K)
phonon_scattering_time = 1e-12  # Phonon scattering time in seconds
temperature = 300  # Temperature in Kelvin
velocity = 1e4  # Average phonon velocity in m/s

# Calculate phonon dispersion
phonon_frequency = phonon_dispersion_relation(wavevector,
    lattice_constant, omega_zero)

# Calculate thermal conductivity
thermal_conductivity =
    calculate_thermal_conductivity(mean_free_path, specific_heat,
    velocity)

# Calculate carrier mobility
carrier_mobility =
    calculate_carrier_mobility(phonon_scattering_time, temperature)

# Output results
print("Phonon Frequency:", phonon_frequency, "Hz")
print("Thermal Conductivity:", thermal_conductivity, "W/(m*K)")
print("Carrier Mobility:", carrier_mobility, "m^2/(V*s)")
```

This code defines three functions:

- `phonon_dispersion_relation` calculates the phonon frequency based on the wavevector and base frequency.
- `calculate_thermal_conductivity` computes the thermal conductivity of the semiconductor using the mean free path, specific heat, and phonon velocity.
- `calculate_carrier_mobility` calculates the carrier mobility influenced by phonon scattering time and temperature.

The provided example computes the phonon frequency, thermal conductivity, and carrier mobility in a semiconductor context, then prints the resulting values.

Chapter 31

Spintronics

The field of spintronics explores the role of electron spin in semiconductors and its potential for novel electronic devices. In this chapter, we provide an introduction to the fundamentals of spintronics, including the concept of electron spin, spin-dependent transport phenomena, and the potential applications of spin-based devices.

Electron Spin

The concept of electron spin lies at the heart of spintronics. In quantum mechanics, the spin of an electron is an intrinsic property that arises from its quantum mechanical nature. Unlike classical angular momentum, electron spin is quantized in discrete values of $\pm\frac{1}{2}\hbar$, where $\hbar$ is the reduced Planck's constant. This property gives rise to two distinct spin states, often denoted as $\uparrow$ and $\downarrow$. The ability to manipulate and control electron spin opens up new possibilities for information storage and processing.

Spin-Orbit Interaction

Spin-orbit interaction is a fundamental phenomenon in spintronics that arises from the coupling of an electron's spin and its orbital motion. This interaction influences the behavior of spin in the presence of an electric field or a magnetic field gradient, leading to spin precession and spin relaxation. The spin-orbit interaction plays a crucial role in spin-based devices and can be tailored in different semiconductor materials by engineering their crystal structures and electronic properties.

Spin-Dependent Transport

Spin-dependent transport refers to the transport of electrons with a specific spin orientation in a semiconductor material. Spin-dependent phenomena include spin filtering, where only electrons with a particular spin direction can pass through specific barriers or interfaces, and spin polarization, where the spin orientation of carriers deviates from thermal equilibrium due to external influences. The understanding and manipulation of spin-dependent transport are essential in the development of spintronic devices.

Spin Relaxation

Spin relaxation, also known as spin decoherence, is a process that causes the loss of electron spin information over time. Several mechanisms contribute to spin relaxation, including spin-orbit interaction, phonon scattering, and magnetic impurities. By studying and controlling spin relaxation processes, researchers aim to extend the coherence time of electron spins, enabling more robust and efficient spin-based devices.

Spin Hall Effect

The spin Hall effect refers to the generation of a transverse spin current in a semiconductor material when an electric current is applied across it. This effect arises from spin-orbit interaction and can be utilized to separate spin-up and spin-down electrons, leading to the generation of spin currents without the need for ferromagnetic materials. The spin Hall effect has gained significant attention in spintronics research due to its potential for spin manipulation and efficient spintronic device design.

Spin-Based Devices

Spin-based devices encompass a wide range of devices that utilize electron spin for information storage, manipulation, and transport.

Examples include spin valves, where the resistance depends on the relative orientation of two ferromagnetic layers, and spin transistors, where the control of spin currents enables efficient switching. Spin-based devices have the potential to revolutionize electronics by offering faster, smaller, and more energy-efficient technologies.

Spin-Orbit Torque

Spin-orbit torque is a phenomenon that arises when an electric current induces a transfer of spin angular momentum to a magnetization vector in a magnetic material. This torque effect allows for efficient manipulation of magnetization, enabling the writing and reading of information in magnetic storage devices. Spin-orbit torque provides an alternative to traditional magnetoelectric devices and has promising applications in a variety of spintronic devices.

Magnetization Dynamics

Magnetization dynamics is the study of how magnetic moments respond to external influences, such as magnetic fields or spin currents. Understanding the behavior of magnetization is crucial for the design and optimization of spin-based devices. The dynamics of magnetization involves phenomena such as precession, damping, and domain wall motion, which can be described by micromagnetic simulations and various mathematical models.

Spin-Photon Interaction

Spin-photon interaction explores the interaction between spin systems and electromagnetic radiation. In spintronics, photons are used for spin manipulation, detection, and communication. Techniques such as optical spin injection, spin resonance, and optically detected magnetic resonance enable the efficient transfer of spin information between electron spins and photons. Spin-photon interaction opens up new avenues for spin-based quantum technologies and quantum information processing.

In this chapter, we provided an introduction to the role of electron spin in semiconductors. We discussed the concept of electron spin and its quantized nature. The spin-orbit interaction and its influence on spin behavior were explored, along with spin-dependent transport phenomena and the spin relaxation process. The spin Hall effect and its potential for spin manipulation were discussed. Additionally, we introduced various spin-based devices, such as spin valves and spin transistors, highlighting their potential applications and advantages. The spin-orbit torque and its role in magnetization manipulation were presented, as well as the dynamics of magnetization. Finally, the spin-photon interaction and its implications for spin-based quantum technologies were discussed."'latex

Python Code Snippet

Below is a Python code snippet that implements important equations and algorithms related to spintronics, including the calculation of spin relaxation time, spin Hall effect, and magnetization dynamics.

```python
import numpy as np

def calculate_spin_relaxation_time(mechanical_viscosity,
    spin_diffusion_coefficient):
    '''
    Calculate the spin relaxation time based on mechanical viscosity
    and spin diffusion coefficient.
    :param mechanical_viscosity: Mechanical viscosity of the
    material in kg/(m·s).
    :param spin_diffusion_coefficient: Spin diffusion coefficient in
    m²/s.
    :return: Spin relaxation time in seconds.
    '''
    return mechanical_viscosity / (spin_diffusion_coefficient *
    np.power(np.pi, 2))


def calculate_spin_hall_current(current_density,
    spin_momentum_conversion_factor):
    '''
    Calculate the transverse spin current generated by the Spin Hall
    effect.
    :param current_density: The applied current density in A/m².
    :param spin_momentum_conversion_factor: Spin momentum conversion
    factor (dimensionless).
    :return: Transverse spin current density in A/m².
    '''
    return spin_momentum_conversion_factor * current_density
```

```python
def magnetization_dynamics(magnetization, damping_factor,
    external_field, time):
    '''
    Calculate the response of magnetization to external influence
        over time.
    :param magnetization: Initial magnetization vector (numpy
        array).
    :param damping_factor: Damping factor (dimensionless).
    :param external_field: External magnetic field vector (numpy
        array).
    :param time: Time in seconds over which to calculate dynamics.
    :return: Final magnetization vector after time t.
    '''
    gamma = 2.21e5  # gyromagnetic ratio in rad/(s·T)
    precession = np.cross(magnetization, external_field) * gamma
    damping = -damping_factor * magnetization

    final_magnetization = magnetization + (precession + damping) *
        time
    return final_magnetization


# Inputs for the calculations
mechanical_viscosity = 0.001  # kg/(m·s)
spin_diffusion_coefficient = 1e-5  # m²/s
current_density = 1e6  # A/m²
spin_momentum_conversion_factor = 0.3  # dimensionless
initial_magnetization = np.array([1.0, 0.0, 0.0])  # Example
    magnetization vector
damping_factor = 0.1  # dimensionless
external_field = np.array([0.0, 0.0, 1.0])  # Applying external
    field in z-direction
time = 0.01  # seconds

# Perform calculations
spin_relaxation_time =
    calculate_spin_relaxation_time(mechanical_viscosity,
    spin_diffusion_coefficient)
transverse_spin_current =
    calculate_spin_hall_current(current_density,
    spin_momentum_conversion_factor)
final_magnetization = magnetization_dynamics(initial_magnetization,
    damping_factor, external_field, time)

# Output results
print("Spin Relaxation Time:", spin_relaxation_time, "seconds")
print("Transverse Spin Current Density:", transverse_spin_current,
    "A/m²")
print("Final Magnetization Vector:", final_magnetization)
```

This code defines three functions:

- `calculate_spin_relaxation_time` computes the spin relaxation time based on mechanical viscosity and spin diffusion coefficient.
- `calculate_spin_hall_current` calculates the transverse spin current generated by the Spin Hall effect based on the current density and conversion factor.
- `magnetization_dynamics` models the response of magnetization to an external magnetic field over time, accounting for precession and damping.

The provided example calculates the spin relaxation time, transverse spin current density, and the final magnetization vector after a given time period, then prints the results. "'

Chapter 32

Optoelectronics

In this chapter, we delve into the fascinating field of optoelectronics, which revolves around the study of light-matter interaction in semiconductors. The ability of semiconductors to absorb, emit, and manipulate light forms the basis for numerous optoelectronic devices, such as light-emitting diodes (LEDs), photodetectors, and solar cells. As a mathematics PhD, we will explore the mathematical models used to describe the behavior of optoelectronic devices, delving into the underlying physics behind light-matter interaction.

Optical Absorption

Optical absorption is a fundamental process in optoelectronics, where a semiconductor material absorbs photons and converts their energy into electron-hole pairs. The absorption coefficient, denoted by α, characterizes the rate at which light is absorbed as it propagates through the material. Mathematically, the relationship between the intensity of incident light $I_0(z)$ and the intensity of transmitted light $I(z)$ can be described using the Beer-Lambert law:

$$I(z) = I_0(z) \cdot e^{-\alpha z} \tag{32.1}$$

where z is the thickness of the semiconductor sample.

Direct and Indirect Band Gap Semiconductors

Semiconductors can be classified into two categories based on their band structure: direct band gap and indirect band gap materials. In direct band gap semiconductors, the minimum energy of the conduction band and the maximum energy of the valence band occur at the same momentum in the Brillouin zone, allowing for efficient emission and absorption of photons. On the other hand, indirect band gap materials have a different momentum for the minimum of the conduction band and the maximum of the valence band, resulting in a lower probability of radiative transitions. This distinction has significant implications for the design and efficiency of optoelectronic devices.

Recombination and Generation Rates

Recombination and generation rates in semiconductors play a crucial role in determining the efficiency of optoelectronic devices. Recombination refers to the process by which electrons and holes recombine, resulting in the emission of photons or non-radiative processes. Generation, on the other hand, denotes the optical excitation of electron-hole pairs, often due to incident photons or other external stimuli. The recombination and generation rates can be quantified using mathematical models based on Shockley-Read-Hall (SRH) theory or Auger recombination for nonradiative processes.

Radiative and Non-Radiative Recombination

In optoelectronic devices, recombination processes can be classified as either radiative or non-radiative, depending on whether they result in the emission of photons. Radiative recombination occurs when an electron and hole recombine, emitting a photon with energy equal to the band gap energy. This process is highly desirable for light-emitting devices like LEDs. In contrast, non-radiative recombination involves electron-hole recombination without photon emission, often resulting in energy dissipation as heat. Non-radiative recombination can significantly reduce the efficiency of optoelectronic devices.

Luminescence and Electroluminescence

Luminescence refers to the emission of light from a material, typically as a result of recombination processes. Photoluminescence, a type of luminescence, occurs when a material is excited by incident photons, and recombination subsequently emits photons with a lower energy. Photoluminescence is commonly used to characterize the optical properties of semiconductors. Electroluminescence, on the other hand, is the emission of light occurring directly as a consequence of an electric current passing through a device, as observed in LEDs. Modeling the luminescence and electroluminescence processes involves understanding the carrier recombination and quantum efficiency of the device.

Theoretical Models for Light-Matter Interaction

The interaction of light with matter in optoelectronic devices can be described by theoretical models such as the semiconductor Bloch equations (SBE), rate equations, or density matrix formalism. These models involve a set of coupled differential equations that describe the dynamics of carriers, photons, and their interactions. The solution of these equations provides insight into the behavior and performance of optoelectronic devices, allowing for the optimization of design parameters to enhance device efficiency.

Photonic Structures

Photonic structures, such as optical cavities and photonic crystals, play a vital role in controlling and manipulating light in optoelectronic devices. Optical cavities, consisting of materials with high refractive index contrast, confine light within a small volume, enhancing light-matter interaction and enabling efficient emission or absorption. Photonic crystals, on the other hand, are periodic structures that exhibit specific band gaps for certain wavelengths, allowing for the control of light propagation and modulation. Modeling the optical properties of such structures requires sophisticated techniques such as finite-difference time-domain or transfer matrix methods.

Optical Waveguides

Optical waveguides are essential components in optoelectronic devices for the efficient transmission and routing of light. These waveguides confine and guide light along a specific path, typically using materials with different refractive indices. The waveguide structure and the refractive index profile impact the mode properties, dispersion, and losses in the waveguide. Modeling waveguide behavior involves solving Maxwell's equations and applying appropriate boundary conditions, allowing for the analysis of propagation characteristics and loss mechanisms.

Photodetectors

Photodetectors are optoelectronic devices that convert incident light into an electrical signal. Modeling the behavior of photodetectors involves understanding the absorption of photons, generation of electron-hole pairs, and subsequent carrier transport and collection. The responsivity, quantum efficiency, and noise characteristics are crucial parameters for evaluating photodetector performance. Advanced models, such as the drift-diffusion equation coupled with the equations for optical absorption, can adequately describe the behavior of photodetectors.

Solar Cells

Solar cells are devices that convert sunlight directly into electricity through the photovoltaic effect. Modeling solar cells aims to optimize their performance by understanding factors such as carrier generation, transport, and collection, as well as factors that may limit their efficiency, including recombination losses and optical losses. Various mathematical approaches, such as the detailed balance model, Shockley-Queisser limit, and drift-diffusion equations, can be utilized to study and improve solar cell efficiency.

In this chapter, we explored the field of optoelectronics, focusing on the mathematical models used to describe light-matter interaction in semiconductors. We discussed optical absorption, direct and indirect band gap semiconductors, recombination and generation rates, radiative and non-radiative recombination, luminescence and electroluminescence, as well as theoretical models for light-matter interaction. Additionally, we touched on photonic structures, optical waveguides, photodetectors, and solar cells, highlighting their significance in optoelectronic device design and performance. By understanding the mathematical foundations of optoelectronics, we can advance the development of more efficient and impactful optoelectronic technologies.

Python Code Snippet

Below is a Python code snippet that implements important equations and algorithms mentioned in the chapter on optoelectronics, particularly focusing on optical absorption, carrier generation and recombination rates, as well as calculations related to photodetector and solar cell characteristics.

```python
import numpy as np

def calculate_optical_absorption(I0, alpha, z):
    '''
    Calculate the transmitted light intensity after absorption in a
    ↪ semiconductor.
    :param I0: Intensity of incident light (W/m^2).
    :param alpha: Absorption coefficient (m^-1).
    :param z: Thickness of the semiconductor sample (m).
    :return: Transmitted light intensity (W/m^2).
    '''
    I = I0 * np.exp(-alpha * z)
    return I


def calculate_recombination_rate(np, tau):
    '''
    Calculate the radiative recombination rate in semiconductors.
    :param np: Product of electron and hole concentration (n * p) in
    ↪ m^-3.
    :param tau: Carrier lifetime (seconds).
    :return: Recombination rate (m^-3/s).
    '''
    A = 1e-12  # Radiative recombination coefficient (m^3/s)
    R_rad = A * np / tau
```

```python
    return R_rad


def calculate_quantum_efficiency(photon_generation_rate,
    recombination_rate):
    '''
    Calculate the quantum efficiency of an optoelectronic device.
    :param photon_generation_rate: Rate at which photons are
        generated (photons/s).
    :param recombination_rate: Rate at which carriers recombine
        (m^-3/s).
    :return: Quantum efficiency (dimensionless).
    '''
    QE = photon_generation_rate / (photon_generation_rate +
        recombination_rate)
    return QE


def calculate_photodetector_parameters(P_in, R):
    '''
    Calculate the output current of a photodetector based on
        incident power and responsivity.
    :param P_in: Incident light power (W).
    :param R: Responsivity of the photodetector (A/W).
    :return: Output current (A).
    '''
    I_out = R * P_in
    return I_out


def calculate_solar_cell_efficiency(V_oc, I_sc, P_in):
    '''
    Calculate the efficiency of a solar cell.
    :param V_oc: Open circuit voltage (V).
    :param I_sc: Short circuit current (A).
    :param P_in: Incident solar power (W).
    :return: Efficiency (percentage).
    '''
    P_out = V_oc * I_sc        # Output power
    efficiency = (P_out / P_in) * 100    # Efficiency in percentage
    return efficiency


# Inputs for the calculations
I0 = 1000    # Intensity of incident light (W/m^2)
alpha = 0.1    # Absorption coefficient (m^-1)
z = 0.01    # Thickness of semiconductor (m)

n = 1e21    # Electron concentration (m^-3)
p = 1e21    # Hole concentration (m^-3)
tau = 1e-6    # Carrier lifetime (s)
```

```python
photon_generation_rate = 1e20  # Rate of photon generation
↪ (photons/s)
P_in = 0.002  # Incident power in a photodetector (W)
R = 0.5  # Responsivity of the photodetector (A/W)

V_oc = 0.5  # Open circuit voltage (V)
I_sc = 0.03  # Short circuit current (A)
P_in_solar = 1  # Incident solar power (W)

# Calculations
transmitted_intensity = calculate_optical_absorption(I0, alpha, z)
np_product = n * p
recombination_rate = calculate_recombination_rate(np_product, tau)
quantum_efficiency =
↪ calculate_quantum_efficiency(photon_generation_rate,
↪ recombination_rate)
photodetector_current = calculate_photodetector_parameters(P_in, R)
solar_cell_efficiency = calculate_solar_cell_efficiency(V_oc, I_sc,
↪ P_in_solar)

# Output results
print("Transmitted Intensity:", transmitted_intensity, "W/m^2")
print("Recombination Rate:", recombination_rate, "m^-3/s")
print("Quantum Efficiency:", quantum_efficiency)
print("Photodetector Output Current:", photodetector_current, "A")
print("Solar Cell Efficiency:", solar_cell_efficiency, "%")
```

This code defines several functions:

- `calculate_optical_absorption` computes the transmitted light intensity after absorption in a semiconductor.
- `calculate_recombination_rate` calculates the radiative recombination rate based on electron and hole concentration and carrier lifetime.
- `calculate_quantum_efficiency` evaluates the quantum efficiency of an optoelectronic device.
- `calculate_photodetector_parameters` computes the output current of a photodetector based on incident power and responsivity.
- `calculate_solar_cell_efficiency` determines the efficiency of a solar cell given its output parameters.

The provided example runs calculations for optical absorption, recombination rates, quantum efficiency, photodetector output current, and solar cell efficiency, then prints the results.

Chapter 33

Advanced Computational Methods

In this chapter, we explore advanced computational methods that play a crucial role in modeling semiconductor physics. These techniques allow us to numerically solve complex mathematical equations that describe the behavior of semiconductors and their devices. One such method we delve into is the Finite Element Method (FEM), which is widely used for solving partial differential equations (PDEs) arising in semiconductor physics. Additionally, we touch upon other numerical approaches that are relevant for simulating and analyzing semiconductor devices.

Finite Element Method (FEM)

The Finite Element Method (FEM) is a numerical technique used to discretize a continuous problem into smaller, simpler subproblems. It is particularly useful for solving PDEs, including the Poisson equation and the drift-diffusion equation, which play a central role in modeling semiconductor behavior. FEM divides the computational domain into a collection of finite elements, each represented by a set of nodes. Mathematical functions, called shape functions, are employed to interpolate the unknown variables within each element. By assembling the elemental equations, a system of algebraic equations is formed, which can be solved to obtain the desired solution.

For example, in semiconductor device simulations, FEM can be used to solve the Poisson equation, which describes the electrostatic potential distribution within the device. In two dimensions, the Poisson equation can be written as:

$$\nabla \cdot (\varepsilon \nabla \phi) = -\rho \tag{33.1}$$

where ε is the permittivity, ϕ is the electrostatic potential, and ρ is the charge density. FEM discretizes the domain into small triangular or quadrilateral elements, and the potential is approximated using a set of shape functions. The resulting system of algebraic equations can then be solved using iterative methods or direct solvers.

Mesh Generation

In FEM, the accuracy of the solution depends on the quality of the mesh, which represents the computational domain. Mesh generation involves dividing the physical domain into a collection of elements that approximate its shape. In semiconductor physics, devices often possess complex geometries, making mesh generation a non-trivial task. Various techniques, such as Delaunay triangulation and advancing front methods, can be employed to generate high-quality meshes. Adaptive mesh refinement, where the mesh is refined in regions with steep gradients or high field variations, is also useful for achieving accurate solutions efficiently.

Time-Dependent Simulations

Many semiconductor phenomena involve transient behaviors that evolve over time, such as carrier transport and recombination. Numerical simulations of time-dependent phenomena are crucial for understanding device dynamics and performance. In FEM, time-dependent simulations are often accomplished by employing transient versions of the governing equations, such as the time-dependent drift-diffusion equation. These equations are discretized in both space and time, resulting in a time marching scheme that advances the solution from one time step to the next. Techniques such as implicit or explicit time integration methods, such as the

Crank-Nicolson method or the forward Euler method, can be utilized depending on the nature of the problem.

Carrier Transport Models

Accurate modeling of carrier transport is essential for understanding the behavior of semiconductor devices. Various models, such as the drift-diffusion model, hydrodynamic model, or semiclassical Boltzmann transport equation, can be employed to describe carrier motion. In FEM, the drift-diffusion model is often the method of choice due to its simplicity and ability to handle diverse device geometries. The drift-diffusion equations describe the fluid-like motion of charge carriers by combining the continuity equations for carriers with the drift and diffusion terms. The resulting equations can then be discretized using FEM and solved to obtain the carrier distribution and device characteristics.

Schrodinger-Poisson Solver

In some semiconductor devices, particularly those with nanostructures or quantum confinement effects, the Schrodinger equation needs to be solved simultaneously with the Poisson equation. This coupled system, often referred to as the Schrodinger-Poisson solver, accounts for quantum mechanical effects on carrier distribution. One common approach is to decouple the equations and employ an iterative scheme, where the carrier density from the Poisson equation is used as an input for solving the Schrodinger equation, and vice versa. FEM can be utilized to discretize the coupled system, resulting in a large-scale eigenvalue problem that needs to be solved iteratively until convergence.

Other Numerical Approaches

Beyond the FEM, other numerical methods find applications in the field of semiconductor physics. Finite Difference (FD) methods, for example, approximate derivatives using finite difference approximations on a grid. The FD method is relatively simple to implement and can be especially useful for solving PDEs with irregular geometries. Moreover, spectral methods, such as the Fourier transform or Chebyshev polynomials, offer high accuracy and rapid convergence rates for problems with smooth solutions. These methods are particularly suitable for periodic structures or problems requiring high precision.

Software and Simulation Tools

Several software packages and simulation tools have been developed to facilitate the numerical modeling of semiconductor devices. These tools often incorporate FEM or other numerical methods, along with pre- and post-processing capabilities. Examples include COMSOL Multiphysics, Silvaco TCAD, and Sentaurus Device. These tools provide user-friendly interfaces for device simulation, allowing researchers and engineers to explore device characteristics, optimize device designs, and predict device performance. Many of these software packages also offer built-in libraries for modeling carrier transport, quantum effects, and electro-optical properties.

Parallel Computing

Simulating semiconductor devices and phenomena often requires solving large systems of equations, which can be computationally demanding. Parallel computing techniques are employed to distribute the computational workload across multiple processors or compute nodes, thereby reducing simulation time and enabling the solution of larger and more complex problems. Techniques such as domain decomposition, where the computational domain is divided into subdomains that are solved simultaneously on different processors, or message passing interface (MPI) for inter-process communication, are commonly used in semiconductor device simulations. High-performance computing clusters and supercomputers are utilized to take advantage of parallel computing capabilities.

Model Validation and Calibration

When employing advanced computational methods, model validation and calibration are crucial for ensuring the accuracy and reliability of simulation results. Experimental measurements of relevant device characteristics, such as current-voltage characteristics or electro-optical properties, can be used to validate and calibrate the numerical models. Quantitative comparisons between simulated and experimental data are performed, and adjustments are made to the model parameters or assumptions until a satisfactory agreement is achieved. Additionally, sensitivity analysis techniques can be applied to assess the impact of model uncertainties and parameter variations on the simulation results.

In this chapter, we have explored advanced computational methods used in semiconductor physics. We began by introducing the Finite Element Method (FEM) as a versatile numerical technique for solving semiconductor PDEs. We discussed the importance of mesh generation, especially for devices with complex geometries, and the challenges of time-dependent simulations. Moreover, we touched upon carrier transport models, the Schrodinger-Poisson solver for quantum effects, and other numerical approaches such as Finite Difference and spectral methods. We highlighted the availability of software and simulation tools tailored for semiconductor device modeling and the role of parallel computing in handling large-scale simulations. Finally, we emphasized the significance of model validation and calibration for ensuring the accuracy and reliability of numerical results.

By employing these advanced computational methods, researchers and engineers can gain insight into the behavior of semiconductor devices, optimize their design, and make informed decisions for improving device performance.

Python Code Snippet

Below is a Python code snippet that implements important equations and algorithms mentioned in the chapter, particularly focusing on solving the Poisson equation using the Finite Element Method (FEM) for semiconductor simulations, as well as a simple implementation for carrier transport modeling.

```python
import numpy as np
import matplotlib.pyplot as plt
from scipy.sparse import diags
from scipy.sparse.linalg import spsolve

def solve_poisson(epsilon, rho, dx):
    '''
    Solve the Poisson equation using a finite difference method
    ↪ approach.
    :param epsilon: Permittivity of the material.
```

```
    :param rho: Charge density as a 1D numpy array.
    :param dx: Grid spacing.
    :return: Electrostatic potential as a 1D numpy array.
    '''
    n = len(rho)
    # Create the coefficient matrix
    diagonals = [-2 * np.ones(n), np.ones(n - 1), np.ones(n - 1)]
    offsets = [0, -1, 1]
    A = diags(diagonals, offsets).toarray()

    # Right-hand side vector
    b = -rho / epsilon
    b[0] = 0   # Boundary condition at left end
    b[-1] = 0  # Boundary condition at right end

    # Solve the system of equations
    phi = spsolve(A, b)
    return phi

def visualize_potential(x, phi):
    '''
    Visualize the electrostatic potential.
    :param x: Spatial domain.
    :param phi: Electrostatic potential.
    '''
    plt.plot(x, phi)
    plt.title("Electrostatic Potential")
    plt.xlabel("Position (m)")
    plt.ylabel("Potential (V)")
    plt.grid()
    plt.show()

def drift_diffusion_model(n, dx, dt, D, mu, rho, total_time):
    '''
    Simple implementation of the drift-diffusion model for carrier
    ↪ transport.
    :param n: Number of grid points.
    :param dx: Grid spacing.
    :param dt: Time step.
    :param D: Diffusion coefficient.
    :param mu: Mobility of carriers.
    :param rho: Initial charge carrier density as a numpy array.
    :param total_time: Total time for simulation.
    :return: Time evolution of carrier density.
    '''
    time_steps = int(total_time / dt)
    density = np.zeros((time_steps, n))
    density[0, :] = rho

    for t in range(1, time_steps):
        for i in range(1, n - 1):
            # Drift term
```

```python
            drift_term = -(mu * density[t - 1, i] * (density[t - 1,
            ↪       i + 1] - density[t - 1, i - 1]) / (2 * dx))
            # Diffusion term
            diffusion_term = D * (density[t - 1, i + 1] - 2 *
            ↪       density[t - 1, i] + density[t - 1, i - 1]) / (dx **
            ↪       2)
            # Update carrier density
            density[t, i] = density[t - 1, i] + dt * (drift_term +
            ↪       diffusion_term)

    return density

# Inputs
epsilon = 8.854e-12  # Permittivity of free space in F/m
rho = np.array([1e18] * 100)  # Uniform charge density in m^-3
dx = 1e-8  # Spatial grid size in meters
x = np.linspace(0, 99 * dx, 100)  # Position array

# Solve Poisson equation
phi = solve_poisson(epsilon, rho, dx)

# Visualize the electrostatic potential
visualize_potential(x, phi)

# Parameters for carrier transport modeling
n_points = 100
dt = 1e-12  # Time step in seconds
D = 1e-4  # Diffusion coefficient in m^2/s
mu = 1e-2  # Mobility in m^2/(V*s)
initial_density = np.array([1e10] * n_points)  # Initial carrier
↪       density

# Simulate carrier transport
total_simulation_time = 1e-6  # Total simulation time in seconds
carrier_density_evolution = drift_diffusion_model(n_points, dx, dt,
↪       D, mu, initial_density, total_simulation_time)

# Visualization of the carrier density over time
plt.imshow(carrier_density_evolution, aspect='auto', extent=[0,
↪       (n_points-1) * dx, total_simulation_time, 0], cmap='hot')
plt.colorbar(label='Carrier Density (m^-3)')
plt.title('Carrier Density Evolution')
plt.xlabel('Position (m)')
plt.ylabel('Time (s)')
plt.show()
```

This code defines several functions:

- `solve_poisson` solves the Poisson equation using a finite difference method based on the specified permittivity and charge density.

- `visualize_potential` creates a plot of the computed electrostatic potential.
- `drift_diffusion_model` implements a simple drift-diffusion model for carrier transport over time.

The provided example first calculates the electrostatic potential from a uniform charge density, visualizes it, and then simulates the evolution of carrier density in a semiconductor over a specified time duration, resulting in a heatmap representation.

www.ingramcontent.com/pod-product-compliance
Lightning Source LLC
Chambersburg PA
CBHW052155220526
45471CB00004B/1686